HISTORIC SHIP MODELS

HISTORIC SHIP MODELS

of the seventeenth & eighteenth centuries
in the Kriegstein Collection

Arnold and Henry Kriegstein

Copyright © Arnold and Henry Kriegstein 2021

First published in Great Britain in 2021 by
Seaforth Publishing
An imprint of Pen & Sword Books Ltd
47 Church Street, Barnsley
S Yorkshire S70 2AS

www.seaforthpublishing.com
Email info@seaforthpublishing.com

British Library Cataloguing in Publication Data
A CIP data record for this book is available from the British Library

ISBN 978-1-3990-0977-5 (Hardback)
ISBN 978-1-3990-0978-2 (ePub)
ISBN 978-1-3990-0979-9 (Kindle)

All rights reserved. No part of this publication may be reproduced or transmitted in any form or by any means, electronic or mechanical, including photocopying, recording, or any information storage and retrieval system, without prior permission in writing of both the copyright owner and the above publisher.

The right of Arnold and Henry Kriegstein to be identified as the author of this work has been asserted in accordance with the Copyright, Designs and Patents Act 1988

Pen & Sword Books Limited incorporates the imprints of Atlas, Archaeology, Aviation, Discovery, Family History, Fiction, History, Maritime, Military, Military Classics, Politics, Select, Transport, True Crime, Air World, Frontline Publishing, Leo Cooper, Remember When, Seaforth Publishing, The Praetorian Press, Wharncliffe Local History, Wharncliffe Transport, Wharncliffe True Crime and White Owl.

Typeset and designed by Mousemat Design Limited
Printed and bound by Printworks Global Ltd, London & Hong Kong

Dedication

To our parents,
Roman and Cecile, alive in memory

Navy Board model of the *Marlborough* of 1706, built at William Johnson's Blackwall shipyard.

Contents

Acknowledgements		10
Foreword		12
Introduction		13
Chapter 1	**The *Royal James*, 1st rate of 1671** *Lucifer at the helm: terror tactics in the age of sail*	19
Chapter 2	**A Charles II 5th rate c1680** *A 'frigate' ahead of its time/Evolution through natural selection*	39
Chapter 3	**The *Coronation*, 2nd rate of 1685** *England's Great Loss by a Storm of Wind*	49
Chapter 4	**The *Adventure*, 5th rate of 1691** *The 'Glorious Revolution' incites a failed counter-revolution*	61
Chapter 5	**A William III 4th rate c1695** *Identifying ships at sea*	67
Chapter 6	**The *Northumberland*, 3rd rate of 1702** *The line of battle influences ship design*	75
Chapter 7	**The *Marlborough*, 2nd rate of 1706** *Carved and gilded decorations disappear on models and ships*	85
Chapter 8	**The *Diamond*, 4th rate of 1708** *Failure of the Darien Scheme promotes the Union of England and Scotland*	93
Chapter 9	**A Queen Anne 3rd rate c1710** *Launching of men-of-war*	103
Chapter 10	**The *Royal Oak*, 3rd rate of 1713** *Oak leaves conceal a King*	111

Chapter 11	**The *Diamond*, 5th rate of 1723, and the *Greyhound*, 6th rate of 1720** *What to give a prince on his tenth birthday*	123
Chapter 12	**The *Lion*, 4th rate c1738** *The Lion and the young pretender*	131
Chapter 13	**A George II 4th rate c1745** *The establishment stifles innovation*	141
Chapter 14	**The *Namur*, 3rd rate of 1746** *The Namur and the War of Jenkins's Ear*	145
Chapter 15	**A French 64-gun ship c1754, built by Augustin Pic** *Augustin Pic, shipwright and model builder*	151
Chapter 16	**The *Généreux*, 3rd rate of 1785** *Modelling a prize*	163
Chapter 17	**The *Franklin*, American 74-gun ship c1800** *A maquette of an American ship of the line*	169
Chapter 18	**The *Carcass* bomb of 1758** *The Carcass bomb launches a hero's career*	177
Chapter 19	**The *Aetna* bomb of 1776** *Fire and brimstone, wood and canvas*	181
Chapter 20	**The *Sulphur* bomb of 1797** *Copenhagen bombarded*	185
Chapter 21	**A Queen Anne royal barge** *An admiral's barge conveys Sir Cloudesley Shovell from one tragic fate to another*	189
Chapter 22	**An admiral's barge c1710** *The invasion of England is aided by a ship model collection*	195
Chapter 23	**A Georgian admiral's barge c1720** *The elusive craftsmen who converted warships into works of art*	201

CONTENTS

Chapter 24	**An admiral's barge c1775** *Uniforms inform the collector*	207
Chapter 25	**A troop transport c1810** *Redcoats at sea*	211
Chapter 26	**A ship's boat c1750** *A feat of navigation such as the world had never seen*	215
Chapter 27	**A Dutch state yacht c1690** *The hunter of Aardenburg*	219
Chapter 28	**Model figurehead for the *Royal Caroline* 1750** *The master carver at work*	225
Chapter 29	**Model of the *Queen Charlotte* figurehead c1784** *Design of the figurehead for the Queen Charlotte of 1790*	229
Chapter 30	**Model of the foremast of the *Victory* of 1765 with damage sustained at Trafalgar** *Victory at the Battle of Trafalgar*	239
Chapter 31	**Ship models in perspective painted on panels** *An eighteenth-century fashion finds royal favour*	245
Chapter 32	**Photojournalism in the seventeenth century** *Willem van de Velde the elder and younger portray the wooden warship*	253
Chapter 33	**Napoleonic-era ship models** *The ship model trade flourishes during the Napoleonic Wars*	265
Chapter 34	**Care and conservation** *Averting the ravages of dust, light, heat, damp, dryness, insects, trauma and consumption*	275
Chapter 35	**Fakes and forgeries** *The Good, the Bad and the Phony*	276 276
Appendix	**A William III 4th rate c1695**	279
Chapter Notes		283
Index		287

Acknowledgements

Henry and I owe an enormous debt of gratitude to Bob Friedman and Cathy Dupont, our first publishers. Bob was truly the enabler who through dedication, hard work and a seasoned eye was able to turn the dream we had of telling the story of our ship model collection into a beautifully produced book. We could not have had a more gracious and understanding ally, and his wisdom guided every step in the creation and subsequent publication of our first volume. When years went by and our collection grew, we were once again delighted that Bob undertook publication of a second edition, this time under the imprimatur of SeaWatch Books. This was handled with equal aplomb and professionalism and we could not have been happier with the result. Now, as more years have passed, we have once again expanded our collection and revised our story. We are lucky that another publisher, impressed by the production of the first two editions, has agreed to take this on. Despite the passing of time, as we sat down to write the new chapters of our collecting saga we were reminded of the risk Bob took many years ago with two novice authors and a quirky obsession. Thank you, Bob, for your trust and guidance.

This book and its illustrations are the results of the authors' efforts, and blame for errors, mistakes and shortcomings lie squarely on our shoulders. Most of the passages relating to seventeenth-century ships were reviewed by Frank Fox, and we greatly appreciate his comments and suggestions. Grant Walker also helped find mistakes in fact and grammar and his input was most welcomed. Any persistent inaccuracies are the result of our misguided stubbornness.

All the photographs in the volume were taken by ourselves of objects in our collection, except where otherwise noted. Assembling these objects was a very social affair, however, in which numerous individuals participated. Chief among them was our father, Roman, without whom we could not have begun this adventure. His generosity, encouragement and involvement not only resulted in a lifelong pursuit of naval artworks, but more significantly brought us closer together as individuals through a shared passion. Our mother, Cecile, also played no small role, particularly in prodding Roman to buy two of everything, so we twins would not fight. The memories of our collecting experiences are the real treasures this hobby has provided for us.

Fellow collectors are among the most interesting and diverse people who played a role in this enterprise. Some inherited models, others bought them, and all enriched us by their hospitality and enthusiasm. They include Bill Tandler, Jean Bonnaveau, Clarkson Collins, Angus White, Lord and Lady Sandwich, Lord de Saumarez, James Hodgson, David Hansen and Frank Fox. Others remain anonymous. In later years the late Lord Kenneth Thomson entered the field of Admiralty Board ship model collecting, as has his son David, and we found them to be astute and gracious colleagues and enthusiasts.

We have had a great deal of interaction with art dealers and auctioneers, and some have become friends. These include Martyn Gregory, Laurence Langford, Jon Baddeley, Brian Newbury, Michael Florio, Charles Wallrock, Hugh Bett, Glen Mitchell, Charles Miller and Alan and Janice Granby.

Historians and museum curators have played a valued role in sharing their knowledge and perspective over numerous visits and repasts, and their ranks include Grant Walker, Simon Stephens, Frank Fox, John R Stevens, Michael Robinson, David Roberts, Stuart Frank, Donald Canney and Richard Hunter. We must make special note of the late John Franklin, whose knowledge of Admiralty models was prodigious, and exceeded only by his warm, generous and humble spirit. His absence is a constant lament. We are also proud to have known Romola Anderson, who, when we were young, said that we reminded her of her husband (R C Anderson) when he was young.

There has been a surprising lack of interest in Admiralty Board ship models among British art collectors over the time that we have been collecting them, but this is not true of Great Britain's model makers. It has been our privilege to know some of the most gifted and dedicated of these craftsmen, and our knowledge of these models has benefited enormously from our friendship with Donald and Iris McNarry and Philip Wride. Donald was a source of

ACKNOWLEDGEMENTS

encouragement and a fount of information, as well as a link to some of the legendary modellers and historians who passed away before we began collecting. We are indebted to Philip Wride for the condition of many of our models. Philip's skill and abilities rank with the best of the dockyard craftsman of 200 and 300 years ago, and we feel privileged to have had the benefit of his friendship as well as his talents. We are also grateful to Rob Napier for his restoration skills, Antonio Mariani who provided some of the plinths our models rest upon, and Ann and Alan Smith and Neri Zagal, who helped restore the more fragile painted surfaces.

Lastly, our wives and children should be mentioned. While they did not contribute materially to the enterprise, their tolerance (if not encouragement) of the time, attention and travel that this collection diverted from other family activities is supremely appreciated. That our family bonds remain strong is a testament to their forgiving natures.

Henry Kriegstein
Arnold Kriegstein

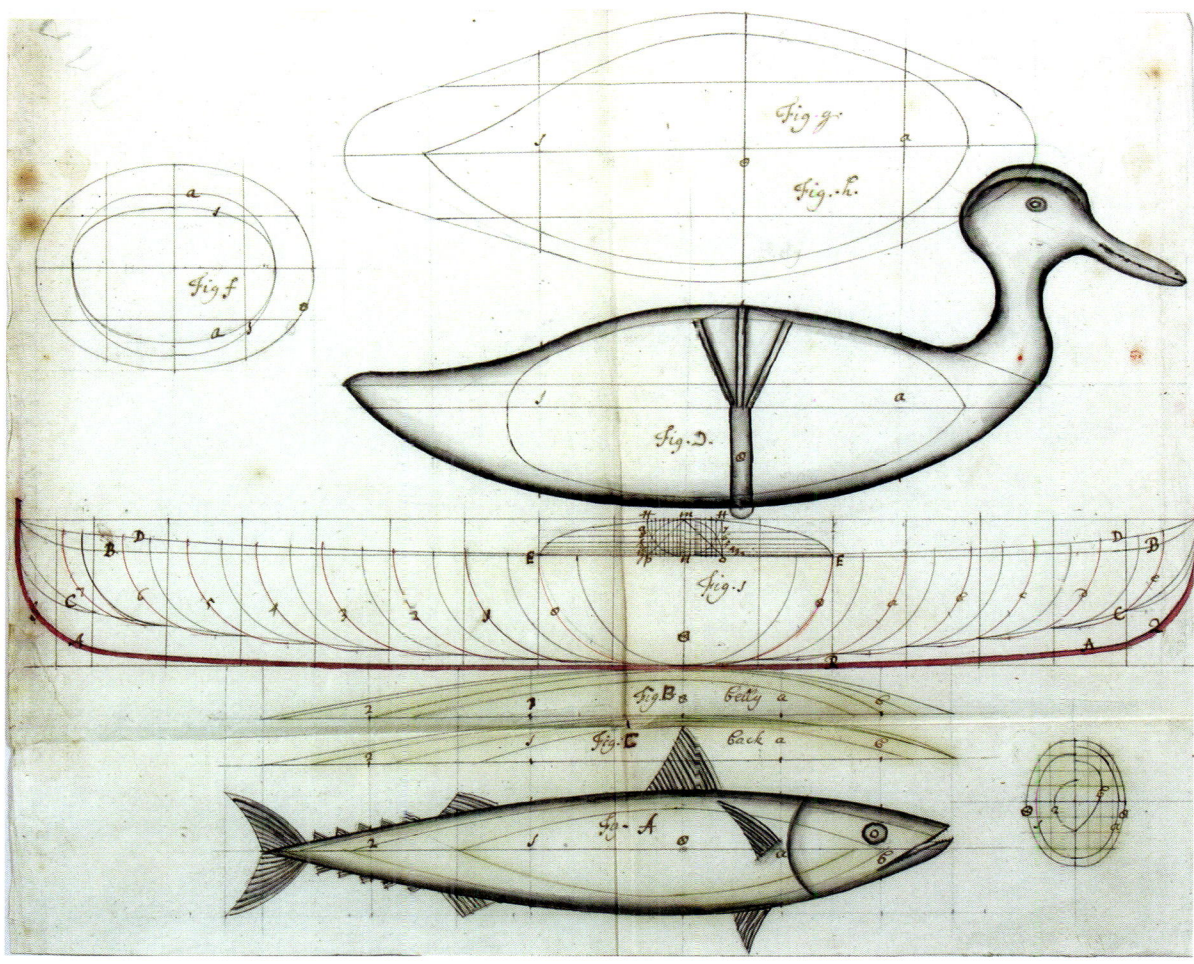

Illustration from an anonymous manuscript titled 'A Short Remark on our Compleat Shipwrights', dated August 1707. It is largely a critique of Edmund Bushnell's treatise *The Compleat Shipwright* first published in London in 1664. With regard to the shape of a ship's hull, the author observes that 'shipwrights and others, for many years … have argued that ships bodies, genuinely formed, should partake of those properties which are incident to nature …' We are told Bushnell 'tells us that the dolphin, salmon, and mackerel dissected and their proportions taken, with all their ellipsis truly squared and cubed, is the true mould for a sailing ship'. Our anonymous author disagrees, and writes: 'Notwithstanding the body of a ship is quite different from the body of such creatures … for a ship must undeniably carry a body in two elements … so that a fowl is much more applicable to such a similitude than a fish can possibly be.' This illustration purports to show how a ship can best be designed by combining the proportions of a mackerel and a duck.

Foreword

FOR OVER FORTY YEARS, I have been in the privileged position of curating one of the most significant collections of ship models in the world. This period also equates to the time that Arnold and Henry have spent assembling what is unquestionably the largest collection of seventeenth- and eighteenth-century ship models in private hands. In addition to the models, their collection also includes oil paintings, ship's plans, books and manuscripts that not only support the wider subject of maritime history, but also focus on the rare references during this period to the models themselves. This has resulted in a truly world-class collection worthy of inclusion in many major museums worldwide.

It still intrigues me that, even in the high-tech world of today, the three-dimensional form of a scale model still proves to be a convincing device to interpret a sometimes complex idea to a general audience, regardless of education, age or gender. With the continuous development of equipment in this digital age, we have been, and are still, drawing upon medical technology to understand how and why these incredibly detailed objects were created. In particular, the use of X-ray, flexible fibre-optical endoscopes and CT scanners has proved an invaluable, non-invasive way of understanding how these models are constructed. In some cases, the virtual and internal exploration of these 'time capsules' has revealed clues as to their makers in the form of pencil markings, signatures and handwritten notes. The models also provide an intriguing view of life below decks with the presence of a variety of fittings such as bilge pumps, capstans, guns, and the layout and decoration of the living accommodation. In addition to illustrating the construction and graceful lines of the hull, the models also exhibit the beautifully executed carved and painted decoration of these warships, an area of interest to both fine art and maritime enthusiasts alike. When one can bring together the traditional research material of archives and images of similar ship model collections via online catalogues, it is not surprising that our understanding of these objects has come a long way in the last ten years.

Arnold and Henry's passion for hunting down these rare masterpieces is still as strong as ever. They are always willing to share the results of their meticulous research and encourage the study of their collection when approached by serious enquirers and model makers. This generosity is a true testament to them and greatly benefits the wider study of these intriguing objects. The inclusion in this volume of recent acquisitions, together with informative text and detailed colour images, is a welcome addition of record for an ever-growing body of work and one that these beautiful objects deserve.

Simon Stephens
Curator, Ship Model Collection
National Maritime Museum
Greenwich, London
November 2020

Introduction

'YOUR LETTER HAS PROVIDED mental interest and stimulus to an old man – I shall be 92 next week and my wife, who is typing this for me will be 85 so please forgive errors.' So ended a letter written to Henry on 15 July 1975 by R C Anderson, the world's leading authority on Admiralty Board ship models. Henry was a medical student in California at the time, and our father had recently purchased the first such model for what was to become a major collection. Dr Anderson was a founder of the Society for Nautical Research, the first editor of the *Nautical Research Journal*, a founder of the British National Maritime Museum (NMM) and author of the 'Bible' of seventeenth-century rigging, *The Rigging of Ships in the Days of the Spritsail Topmast*.[1] He also assembled a superb collection of Admiralty Board models, all of which he generously donated to the NMM. This letter to Henry was written long after he had purchased his last model, but marked the start of our own lifelong collecting passion.

A Tradition of Private Ownership

The collection formed by us, with generous and enthusiastic assistance from our father, Roman, belongs to a long tradition of private ownership of these models. Unlike the dockyard or arsenal models made by shipwrights on the Continent, which were utilitarian and generally remained at the dockyards, the British examples were almost all dispersed to decorate private homes and were the subject of collecting activity right from their origins in the seventeenth century. Samuel Pepys, who was Charles II's Secretary of the Navy Board, left us several observations regarding these little ships as he was not only a famous diarist and observer of life at court, but was also a great collector of Admiralty Board ship models. It is partly through his *Diary* and letters that we know that the models were built by the shipwrights themselves, who gave them away at their discretion. On 30 December 1663, Pepys records a visit to William Coventry, who had been made a Commissioner of the Navy the previous year, and within one month was to become Secretary to the Lord High Admiral. During Pepys' visit, Christopher Pett, who was the master shipwright at Woolwich dockyard, also dropped by, bearing a ship model: '… and I through the garden to Mr Coventry, where I saw Mr Chr' [Christopher] Pett bring him a model, and indeed it is a pretty one, for a New Year's gift – but I think the work not better done than mine.'[2] Pepys is referring to a model that was given to him by Anthony Deane, assistant shipwright at Woolwich, on 29 September 1662. This passage illustrates that British shipwrights gave models to influential members of the Navy establishment, possibly to influence them. By 1716 Navy Board documents show that shipwrights were required to make models of the ships they proposed to build, presumably to demonstrate naval architectural features for review prior to construction of the ships they represented (see Chapter 24). This would explain why so many models were made. The members of the Navy Board were charged with commissioning ships to be built and to assign specific shipwrights to do the work. Throughout most of the seventeenth and early eighteenth centuries, there were at least five royal dockyards and several private ones that competed for commissions to build ships after funds had been allocated by Parliament. The gift of a splendid ship model, whatever its original role as a part of the shipbuilding process, might have served as an inducement to help direct commissions to the giver. It is therefore not surprising that most of these models were given to Lords of the Admiralty who sat on the Admiralty Board and most often to the first Lord of the Admiralty. These ship models can be referred to as either Navy Board or Admiralty Board models, and we have chosen to designate them as Admiralty Board models to highlight their target audience rather than their origins.

It is clear that for most recipients, the models were valued as decorative objects and displayed in public rooms. In this way, disbursement of models was an unofficial part of the production process. Officially, however, the record is nearly mute. It is remarkable that hundreds of such exquisite objects painstakingly created by hand in numerous workshops over nearly two centuries should be virtually undocumented. To this day, only a handful of official references to the building of early models have come to light (see Chapter 23).

To protect the delicate models, elaborate cases were made.

They were among the first glazed cabinets made in England, and surviving early examples are very rare. We are fortunate to own the oldest one known, the elaborate two-part case for the model of a 5th-rate ship from c1680 (Chapter 2). This cabinet closely resembles the display cases that Pepys had designed for his own models (see *Historical Perspective*, Chapter 2).

Admiralty Board models have appeal on multiple levels. First, there is the aesthetic impact derived from the shape – a wonderful blend of complex curves that beautifully marry form with function; as well as the colour scheme – dominated by mellow shades of varnished wood, enhanced by gilded sculptures, and painted with decorative features in black, gold, and red. The use of red paint on the interior surfaces of the gun deck bulwarks was intended to mute the sight of blood and gore during engagements, but the result is nevertheless aesthetically pleasing. Next is the way these models evoke the cultural world in which they were created – the architectural, mythological or political themes that influenced the style and decoration of each model. For example, the Baroque carvings on seventeenth-century models almost always feature bare-breasted females whose appearance in a later age would have been scandalous. In his 1924 book on ship models, R Morton Nance expressed it this way:

> In the ship of a given period we shall then see something that is all of a piece, not only with its land architecture, but with its costume, and its manners … Similar resemblances between ships of the Restoration period, with their flowing curves and elaborate carving, to cavaliers in feathers, lace and periwigs, to the gabled houses that they lived in, and the high-backed chairs that they sat in – even to their flourished handwriting – are obvious enough …[3]

In the seventeenth century, the building and collecting of ship models can be viewed as part of the desire to understand or demystify large and complex parts of the tangible world by altering their scale and 'domesticating' them. This trend was driven by scientific inventions such as the telescope and the Mercator projection. Thus, the orrery allowed the solar system to fit on a tabletop, and maps could depict the world on a sheet of paper. Landscape paintings and miniature portraits became popular, and it is easy to understand the appeal of a model that could fit into a cabinet yet represent the largest and most complicated wooden object ever built.

Then there is the appreciation of the consummate craftsmanship required to construct such a complex object from tiny and intricate pieces, all precisely shaped by hand and long before the advent of power tools. Finally, there is an appeal that stems from the historic importance of these models. They are unique as accurate, contemporary representations of early Royal Naval ships, the immensely important weapons upon which the security and wealth of the English nation depended. On a personal note, we find it impossible to study these models without thinking about the social and historical context from which they emerged. They are portals that allow us to glimpse a long-gone age that can live again only in our imaginations. To convey a sense of this, we have ended each chapter with a description of the associations each model evokes for us.

In the seventeenth century, collections of Admiralty models were formed by King James II (who was Lord High Admiral in 1660–73 and again in 1685), Peter the Great of Russia, Samuel Pepys (who served as Clerk of the Acts in 1660–73 and Secretary of the Admiralty in 1673–79 and again in 1684–89), and a succession of Lords of the Admiralty, members of the Admiralty Board, and other naval administrators. These first collectors were mainly individuals who were in positions of power in the naval administration of the Stuarts, and they established a tradition of private ownership and display of these miniature ships. There is no evidence that they paid for them. There is evidence that over time models were occasionally sold. For example, in his volume titled *Naval Minutes*, Pepys remarks, 'Captain Wentworth did in the year '71 sell a model of a 4th rate ship to Monsieur De Vauvre for 55 guineas.' There is also an intriguing record of a 'model of a ship in a glass case' sold at auction by Mr Christie in Pall Mall, London, on 8 December 1766. It was bought by a Dr Turner for two pounds one shilling. It was not until the late nineteenth century, however, that any significant recycling of models occurred, and a new breed of collector appeared who valued these models as nostalgic relics and historically significant objects. Early collectors included the Englishmen Captain Charles Hoare, Richard Ihlee and Mr Oatway, and in America, T A Howell, Clarkson A Collins, Jr and Irving R Wiles.

After the First World War there was an increase in the number of stately English country homes that were pulled down, and the subsequent dispersion of their contents helped fuel a renaissance of collecting activity. The most prominent and successful of these collectors was shipping magnate Sir James Caird, whose seventy-two models are now in the NMM. The short

INTRODUCTION

Arnold came across this small volume in an antique shop on the Old Brompton Road, London, over forty years ago. Titled *History of the British Navy*, the top slides open to reveal a model of a 1st-rate man-of-war. The Royal Standard indicates a ship of the period of Queen Anne, a date that matches features of the ship and also the age of the binding. We see no reason to doubt that the artefact is likely to be 300 years old. We feel the metaphor is quite apt and could be extended. Historic ship models, such as those described in the pages of the present book, reflect not only the history of the British Navy but the history of Great Britain itself.

list of connoisseurs also includes Dr R C Anderson, who acquired over twenty examples (also now in the NMM), and Robert Spence. But they and their British colleagues were soon eclipsed by wealthier American contemporaries. Junius Morgan Jr, Frederick C Fletcher, J Templeman Coolidge and George F Harding Jr each purchased one or more of these remarkable objects, but by far the most voracious American collector was the Standard Oil magnate, Colonel Henry Huddleston Rogers. He would eventually acquire over forty examples, but his most important acquisition was the purchase, in 1922, of the entire Sergison collection.

Charles Sergison followed in the footsteps of his friend Samuel Pepys as Clerk of the Acts to the Navy Board beginning in November 1689. His tenure lasted until 1719, and during this time he was able to acquire ten fine Admiralty Board models. For nearly 300 years following his death, they remained on his estate in Cuckfield Park, Surrey, until they came to the attention of R C Anderson, who obtained permission to study and restore them in 1913. Anderson, who was a founding member of the Society for Nautical Research, spent weeks at Cuckfield Park identifying, listing and repairing these models. He eventually moved a number of them to his workshop in Southampton, where they remained for several years undergoing more substantial repairs.

One day, without warning, a van arrived with movers authorised to collect and send the models to the United States. Anderson learned that the entire collection had been purchased by Colonel Rogers. This outraged Anderson, but there was no national repository for these models in England at the time, and therefore little he could do but write editorial complaints in *The London Times* and *The Mariner's Mirror*. This he did, and so in Vol. 8 of *The Mariner's Mirror*, 1922, on p. 379, he lamented:

> The vast majority of our Members will, I am sure, be very sorry to learn that the Cuckfield models have been sold to an American collector. It seems a pity that their late owners should have done this without giving their own countrymen a chance to compete for them. Such, however, is apparently the case. No doubt they will be well cared for in their new home, but it can hardly be disputed that they ought to have remained in England.

A similar situation threatened towards the end of the decade when the 'Mercury' collection of models was for sale, and Colonel Rogers was again interested. This was a collection assembled by the Victorian banker Charles Hoare and included about ten Admiralty Board models.[4] The British public was aroused by a publicity campaign that included an article in the *Illustrated London Times* in 1929. Under the headline 'Keep the 'Mercury Collection of Ship-Models in England! Historic Treasures in Peril of Export', the article made the following appeal:

> The finest collection of old ship-models, that in the Nautical Training Ship '*Mercury*,' at Hamble, Southampton, is to be sold to provide an endowment fund for the establishment, and is in danger of being lost to this country if money is not forthcoming within three months to secure it as a national possession. The amount mentioned as required for endowment

purposes is £30,000. The necessary sum could be raised at once by accepting offers from America, but the Trustees are reluctant to allow such a treasure to go abroad. The Prince of Wales has expressed the hope that the collection will be bought for the nation, or at any rate remain in England and not be dispersed.[5]

This time the Society for Nautical Research, with financial assistance from Sir James Caird, intervened and bought the collection in 1929 for a national museum that they hoped to help establish. It would be five years, however, before the Act to establish the National Maritime Museum would be passed by Parliament. An official repository finally opened in Greenwich in 1934, and models have been flowing into its collections ever since.

Most of the models assembled by the collectors mentioned above have found their way into public institutions here and abroad. During the thirty years that we have been collecting, no fewer than seventeen models have been added to museum inventories by gift or purchase (five – Greenwich, one – Science Museum, London, one – Annapolis, one – Bath Maritime Museum, one – Mystic Seaport Museum, eight – Art Gallery of Ontario). A shrinking number remain in private hands, and many of those that do are described in these pages.

Henry once heard an Old Master paintings collector observe that 'every collector has, in his mind, two collections; the collection that he has, and the collection that he could have had'. For us these words certainly ring true. Not included in this book are models that we did not buy, and some that we did, but subsequently sold or gave away. Prominent among the latter category are five builders' models that are now in museums. These include three British Navy Board models: a sectional model of the *Revenge* of 1718, an early frigate on launching ways and a Napoleonic-era troop transport. The latter two we bought from Lord DeSaumarez at Shrubland Hall, and we sold all three to the late Lord Kenneth Thomson. They are now at the Art Gallery of Ontario, in Toronto, Canada. Two other models no longer in our collection are a take-apart model of King George III's yacht *Royal Sovereign* and a large model of a late nineteenth-century French two-decker. Both of these models we donated to the Mariners' Museum in Newport News, Virginia, many years ago.

We feel privileged to own a collection of Admiralty Board models and to participate in the tradition of private ownership and appreciation of these decorative and significant small treasures. This

Roundhouses, seats of ease and quarter galleries are familiar features on models of seventeenth- and eighteenth-century ships. A related feature seldom shown on ship models, however, is the urinal or pissdale placed on the upper deck. Details show the pissdales on the *Diamond* of 1707 (upper left), the *Marlborough* of 1706 (upper right), a Queen Anne 4th rate c1710 (lower left) and the *Royal Oak* of 1713 (lower right). This must have been a great convenience to sailors, who might otherwise have had to leave their stations and go the toilet, and contemporary ship models offer the best evidence for the placement and appearance of these practical fittings. It is interesting that all four of our Queen Anne period models are equipped with them, always located forward in the waist, against the inboard side of the bulwarks. Each model features a different design and shape, evidence that the builders exercised creative licence in their construction.

book is our means of bringing them to the attention of a wider public through images and text.

References

Lavery, Brian and Simon Stephens, *Ship Models: Their Purpose and Development from 1650 to the Present* (London: Philip Wilson, 1995), p. 256.

Walker, Grant H, 'The Henry Huddleston Rogers Collection of Ship Models at the US Naval Academy'. *Nautical Research Journal*, 44, 2 (June 1999): pp. 74–88.

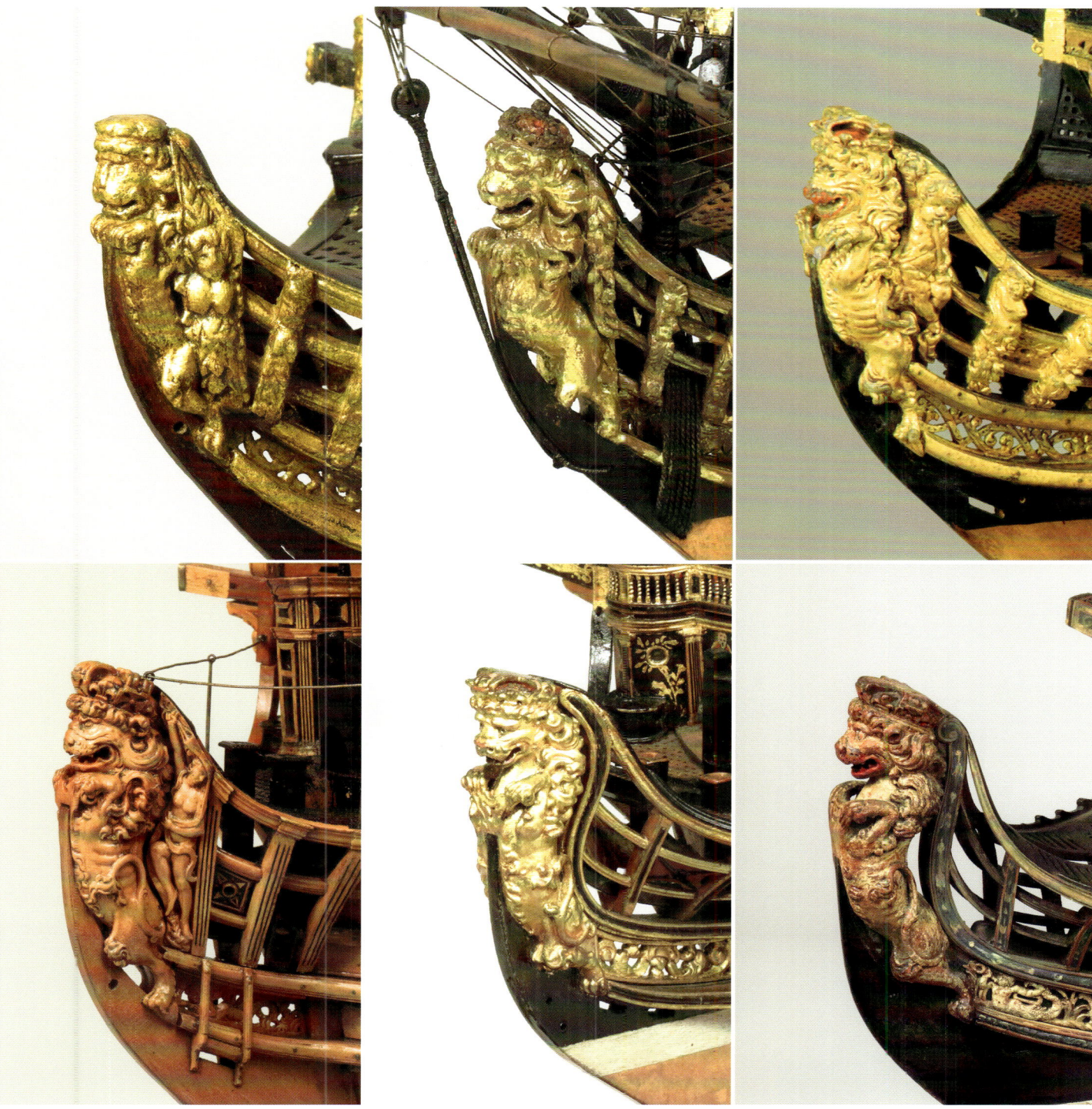

Lion figureheads are depicted here from models arranged left to right from the earliest, a Charles II 5th rate c1680 in the upper left, to the *Namur* of 1746 at lower right. Some lions are solo, others have cherubs or female companions fitted between the beast and the first head timber, as in the earlier examples in the upper row. This practice stopped in the eighteenth century, often leaving an awkward space between the lion's flanks and the hair bracket, as evident in the *Namur*.

Although the royal dockyards were scattered across England, there is a remarkable consistency between the lions carved at the different locations at any given time. They all had the same rampant pose, but as John Franklin pointed out in his book on Navy Board ship models, tails did not appear until the beginning of the eighteenth century, when they suddenly became universal.

CHAPTER 1
The *Royal James*, 1st Rate of 1671

Acquisition

FIRST RATE-SHIPS WERE the largest and most magnificent vessels in the fleet, and we had been haunted by the possibility of acquiring a model of one ever since we came across a photograph of a 1st rate from the Restoration navy that appeared in an issue of the *Mariner's Mirror* in 1912, but whose whereabouts were unknown. For over twenty years we searched for the model referred to as the 'RUSI model' because it was last seen at the Royal United Service Institute museum in 1948. In December 1996 on a brief trip to London, Arnold stopped by the Parker Gallery, one of the few London galleries specialising in naval antiques, to chat with the proprietor, Brian Newbury, about ship models and to check his inventory. It was near closing time on Friday evening, and he had several odd models on display in the window. One was a model of a Second World War destroyer, one resembled the *Mayflower* and one appeared to be an unfinished model of a seventeenth-century warship. This last model had good proportions and a handsome figurehead, and Brian brought it to a table so Arnold could get a better look. Brian called it 'the last of the Parker Gallery's Admiralty Board ship models'. Arnold was amazed, however, at the story Brian told him when asked how he had come to acquire it. His father, Bertram Newbury, had died seven years before. Bertram had been a true connoisseur of ship models and many wonderful and important models had passed through his hands in the heyday of the Parker Gallery. The gallery was London's oldest firm of picture and print dealers, having been founded in 1750. They specialised in naval and military subjects, and often sold models over the years.

When Bertram was literally on his deathbed, he told his son, Brian, that there was a ship model locked away in a cabinet in the basement of the shop. He admonished Brian not to sell this particular model, telling him that the longer he kept it, the more valuable it would become. With Bertram's death the model was duly forgotten. Seven years later Brian noticed a locked chest that had been in the basement for 'donkey's years'. He decided to call a locksmith to have it opened. To his surprise, when the lock gave way and the cabinet doors opened, there was a dusty old ship model inside. He recalled his father's words, put two and two together, and decided that this must be the self-same model. He dusted it off and placed it in the window for sale. This was one week before Arnold stopped by. After relating this story, Brian was ready to close shop, so Arnold took photos of the model and said goodbye. Despite the intriguing story, the model had several parts clearly made of new

Above: This model has the unusual feature, visible here, of narrow gangways on either side of the poop deck leading from the bulkhead stairs approximately halfway to the stern. This provision allows the headroom in the officers' cabins to remain as high as possible.

Left: The *Royal James* has a relatively long beak, as well as considerable sheer at the bow and stern, consistent with its early date. The long gun decks, resulting from the placement of 15 generously spaced guns on the lower deck, produce a low-slung profile uncommon in such a large three-decker.

The bowsprit passes to the starboard of the stem so that the butt end can be stepped alongside the foremast on the lower gun deck. The foremost gun port on the main gun deck is placed between the cathead bracket and the rail, enabling a forward arc of fire. The anchors are stowed, and the anchor buoy can be seen secured to the foremast shrouds.

A lovely feature of this model is the bold acanthus leaf decoration painted along the frieze planking. Circular wreaths around the gun ports were a standard feature at this time, having largely replaced the square port wreaths that were common in pre-restoration ships.

wood as well as some apparently unfinished areas at the stern. Arnold suspected it was a fine copy made in the last century, and he flew home the next day.

Three days later, on Tuesday evening after dinner, Arnold sat down to examine the newly developed photos he had taken on his trip. On top of the pile was a broadside view of the model, but it was underexposed and in near silhouette because the flash on his vintage Leica had failed to fire. The profile seemed oddly familiar. After several seconds Arnold realised that this model looked just like the missing RUSI model, and in the next moment it dawned on him that this *was* the RUSI model! After twenty-five years of searching for a model that was only known from dark black-and-white photos, Arnold had come across it in living colour and not recognised it. The pieces of new wood could now be explained as part of the restoration undertaken by Robert Spence in the 1930s. Arnold immediately grabbed the phone to give Henry the incredible news. We were astounded that the model had finally been found and was for sale, but also concerned because we had not bought it yet! Arnold spent the night pacing sleeplessly until 4am (9am London time), when he called the Parker Gallery. An assistant told him that Brian was not expected for one more hour and kindly offered to have him call when he arrived. Arnold had rarely been more anxious than he was for that long, long, hour. The phone rang promptly at 5am. 'Good morning Brian,' Arnold said, and after a short pause, Brian, sounding quite surprised, replied, 'How did you know it was me?' We laughed at this, but Arnold quickly brought the conversation around to the model. Within twenty-four hours we had not only found, but had purchased the missing *Royal James* model. This was a time for celebration. But there was one more obstacle that could scuttle our hopes of ownership. As we had learned once before (see Chapter 6), buying an important model in England can be much easier than bringing it home. This model would require an export licence.

The Export Licensing Unit of the Department of National Heritage is charged with issuing export licences for works of art of potential cultural, historic or aesthetic importance. The Reviewing Committee on the Export of Works of Art meets regularly to consider arguments for and against export of important objects, taking into account the input from pertinent authorities. The expert for nautical works of art at the time was Simon Stephens of the NMM, and he was consulted in this case. Simon decided that the model merited a closer look, and he made arrangements to inspect the model on the Parker Gallery premises. Once before, we had the export of a model opposed by the NMM, and we knew that the decision could be influenced by how much the model had been altered by restoration. Losses were acceptable, but undetectable alterations would undermine the historical reliability of the model as an authentic example of Restoration warship design. How much of the existing model was original, and how much the work of Robert Spence? To complicate matters, Spence was a master craftsman perfectly capable of making seamless additions or alterations to a model. Proof can be found in the NMM storage depot, where there is a lovely model purchased by the museum at auction in 1944 that was believed at the time to be an authentic seventeenth-century Admiralty Board model of the *St Albans*. However, an identical model also purporting to be the *St Albans* subsequently appeared in the private collection of Robert Spence. In fact, Spence had been the consignor of the *St Albans* model purchased by the NMM. Spence died in 1964, and to this day it is not entirely clear which, if either, of these models is original. Possibly Spence had taken apart one authentic model and reproduced pieces to make two identical models incorporating some of the original in both! It thus came as no surprise to us that Simon planned to subject our *Royal James* model to a thorough inspection, even going so far as to probe the interior with a surgical endoscope.

We had spent over twenty years searching for this model, and while none of our efforts led us to it, we had accumulated some interesting tidbits of information. For example, we had learned that when Spence restored the model in the 1930s he left a note inside detailing what he had done. It now occurred to us that this note could have a bearing on the export licence because it could enhance the historical importance of the model by specifically indicating exactly what had been done to the original. We decided to leave nothing to chance. Arnold made plans for a quick visit to London intending to find the letter before Simon's inspection and read it over. Only then would we know how damaging it might be to our case. To be perfectly honest, the thought of removing the note also crossed our minds, but we resolved not to do that. As it happened, Arnold never found the note, but not for want of trying. He spent two hours on a Saturday morning carefully inspecting and photographing every square inch; even taking the model partially apart, but to no avail. Arnold was very disappointed. He had flown

Nearly all warships in the time of Charles II sported the royal arms beneath the taffrail. In three-deckers, including the *Royal James*, these were enormous carvings that dominated the stern decorations. The coat of arms shown here is a seventeenth-century replacement (the original having vanished long ago) but is an impressive example of miniature carving. The craftsman has made no concession to the small scale and has included every detail that would appear in the full-size carving.

to London to accomplish this one task and had failed. But Arnold was fairly confident that Simon would do no better, and that the letter, if it existed at all, would not be found and would not figure into the export decision. Many months later, however, as Arnold was once again going over the events of that morning, he suddenly realised where Spence may have hidden his note. When Arnold had lifted off the main capstan (a drum-like device used for hauling anchors and shifting guns), he noted that the partners for the capstan spindle, usually two pieces of deck planking with a round hole between them, were in this case built up into a tiny box-like structure. A box! Arnold could think of no reason why this piece should not have been a simple thin plank. The miniature box was only 1in square and ¼in thick, and he decided it must contain the fabled note.

Even after Simon had examined, endoscoped and photographed the model, he was not ready to make a decision on granting the export licence. We had shared with him what we knew of the model's history and provenance, and yet weeks went by without a decision. We began to plan how we would present our case in favour of export if it came to a hearing before the committee. Finally, and not without some additional anxious moments on our part, a decision was made to grant an export licence. We immediately agreed to lend the model to the NMM for an exhibition centred around a conference on the subject of Navy Board models that was scheduled for several weeks later. We have always regarded these models as historical treasures and take very seriously our own responsibility as their custodians. We therefore happily participated in the NMM conference, lending our *Royal James*, *Royal Oak* and *Généreux* models, as well as a painting of a model of the *Royal George* by Joseph Binmer (see Chapter 31).

Long after we had obtained our 'definitive' export licence, we asked Philip Wride, who restored this model for us, to remove the little structure under the capstan and see if he could open it. Sure enough, it was a miniature box with a sliding lid, and inside was the carefully folded handwritten note left by Robert Spence in 1936 with the details of his 'gentle' restoration of the model.

Left: This photograph provides a perspective that one might have had when approaching the stern of the *Royal James* by barge. Stern and quarter galleries were enclosed at this date. Open galleries and balconies are features that would appear only at the end of Charles II's reign. The Jacob's ladders, visible at either side, were used, when needed, to exit to a waiting barge. Two tiers of lights are fitted, and the stern is pierced for a total of 10 guns.

Provenance

Sir Anthony Deane, the original owner of this model, kept it for only a few years until Samuel Pepys persuaded him to give it away. Pepys, the famed seventeenth-century diarist, was a close friend of Deane and was also one of the founders of the Mathematical School at Christ's Hospital. The Mathematical School was established to provide an education in mathematics and navigation for poor children and foundlings in order to prepare them for a career in the Royal Navy. Pepys persuaded Deane to give his ship model to the school to serve as an instructional aid and to promote the naval education of the students. This is recorded in a memo Pepys made to himself many years later. Pepys assiduously compiled notes during his years of service on the Navy Board in preparation for a book on the history of the Navy that he intended to write, but never completed, and his undated memo reads: 'Recollect Sir A[nthony] Deane's and mine own particular acts of service in the Navy, especially about 1675, and particularly Christ's Hospital, where he has given a model and I time, discourse, and instruments.' – S Pepys, *Naval Minutes*.[1] Unfortunately, some ten years later, Pepys came to regret his role in persuading his friend to donate the *Royal James* model. On a visit to the school around 1685 he noted: 'Observe the little use, or rather total neglect, of that admirable model of the Royal James (I think it is) made by Anthony Deane and given by him to Christ's Hospital.' – S Pepys, *Naval Minutes*.[2]

We have found a series of references that document the preservation and display of ship models at the Mathematical School throughout the eighteenth and nineteenth centuries. A bill survives from 1705 recording that John Green was paid £6 'for the new Rigging of the Ship which stands in a case in the Mathematical School', and a shipwright at Woolwich was paid 23 guineas to build a new ship model.[3] When Flamsteed, the master of astronomy, was governor in the early eighteenth century, he admonished that the boys' 'Latin should not grow rusty in the room where stood the models of men-of-war and the terraqueous globe.'[4] In the nineteenth century, Wilson records that, 'In a room between the upper end of the Hall and the Mathematical school are correct models of the various sized vessels now composing the wooden walls of old England, with the name of the donors. These have been given with the view of illustrating the system of mathematics, and are doubtless of great assistance to the Mathematical masters.'[5] The *Royal James* model apparently remained at the school for over 225 years until around 1902, when the school moved out of the City of

London to more spacious and tranquil grounds in Horsham, Sussex. However, the ship models that had been displayed at the school in London were never transferred to the new quarters. A series of sales was held to dispose of books, furnishings, assorted equipment and paintings prior to the move, and this model was presumably included among the de-accessioned items.

At about the time that Christ Hospital was moving out of London, a model with an old label claiming that it represented the *Royal James* was placed on display at the relatively new Royal United Services Institution (RUSI) Museum located in the banqueting hall of the seventeenth-century Palace of Whitehall. This museum served as a repository for a wide variety of interesting and important naval 'relics' at a time before England had a national maritime museum. The model of the *Royal James*, which appeared at the RUSI Museum at the same time as a model of the same ship disappeared from Christ's Hospital, must surely be the same model. It was while the model was exhibited at the RUSI that the first photographs of it appeared in the *Mariner's Mirror*. These photos taken in 1912 document losses to the carved decoration and woodworm damage echoing the state of 'total neglect' lamented by Samuel Pepys over 200 years before, but they also show a fine and graceful hull crowned by a magnificent equestrian figurehead.[6] These photos have been repeatedly reproduced, most recently in Frank Fox's book *Great Ships: the Battlefleet of Charles II*. While it was still at the museum, the model was partly restored by Robert Spence, a noted model craftsman, but no photographs exist showing its restored appearance. The model remained at the RUSI Museum, on exhibit in the painted gallery of Whitehall Palace, until 1948 when the 'rightful owner' contacted the museum and made arrangements to pick it up. It then dropped from public view for the first time in 320 years. The model was to remain hidden for the next half-century, despite concerted efforts to find it by a handful of scholars and determined researchers including ourselves.

~ Description ~

CONDITION

This is model number 13 in Henry Culver's book, *Contemporary Scale Models of Vessels of the 17th Century*. Despite Pepys' lamentation concerning its neglected state, made just ten years after its construction, the model has survived with its most important

This is one of Lord Dartmouth's copies of the manuscript written in 1685 by Edward Battine entitled, *The Method of Building, Rigging, Apparelling and furnishing his Majesties Shipps of Warr according to their Rates …* This example has been bound in roan leather by the 'Naval binder' and retains its original silver gilt clasps. The manuscript contains multiple tables listing particulars of the *Royal James* of 1675, and the detailed measurements of masts, blocks, and rigging were used in restoring the *Royal James* model.

elements preserved. The hull of the model is essentially intact, and it retains key pieces of carved decoration as well as beautiful and well-preserved painted frieze work along the topsides and at the stern. Because many pieces of carved decoration had been lost, and because many of the replacements made by Robert Spence were obvious and crude, we decided to replace the inferior work with better-quality substitutes and to add the obvious missing elements. This could be accomplished with very little conjecture for two

The Weight of the Royʳˡ James with all her Loading	Tuns	C	q͗
The Hull	1249	—	—
Masts, Yards, Tops & Caps	32	16	—
Ballast	145	1	8
Bread for 750 men 6 months	56	15	—
Beef for dᵒ at 4ᵉ	32	2	3·12
Sized Fish 4¼ ᵉ ⎱ each a week	16	1	1·26
Butter 6 ounces ⎰	10	—	3·18
Cheese ¾ ᵉ	3	—	1·2
Pease 1½	6	—	— 21
Beer 1 gallon each a day	349	18	1·2
Water	15	1	1·10
Wood for fireing	26	—	—
Rigging	26	—	—
Cables	38	2	—
Cabletts, hawsers &c	20	12	1·7
Anchors	15	8	—
Guns	178	—	—
Carriages	26	8	—
Shott	55	10	—
Gunners Stores	11	14	—
Powder	20	—	—
Boatsw:ᵣˢ ⎱	9	7	—
Carpenters ⎰ Stores	8	13	—
Captains ⎫	3	—	—
Officers ⎬ luggage	3	—	—
Seamens ⎭	20	—	—
750 men	50	—	—
Blocks	3	6	—
Boats	3	5	—
	2473 Tuñ	6 3·12	

Her draught of water was 20½ abaft & 19½ afore, her body containing about 85000 cube feet, agreeable to the rule aforesᵈ: but nothing of so irregular & so great a bulk can be taken exactly by measᵉ or weight

Battine used the *Royal James* as the prototype 1st rate in his manuscript. From this page one learns that for a three-month voyage, the beer stowed in the hold weighed twice as much as all of the cannon.

related reasons. The hull decoration of ships of this period consisted largely of carved brackets, or caryatids, and all were similar to others at the same level. For example, hermaphrodites with female torsos and male heads usually appear at all the deck bulkheads, while floral decorations or crouching cherubs usually appear between the lights at the stern and quarter galleries. This model had luckily retained at least one example of carved bracket at each level. This allowed correct, reliable replacement of all the missing carvings based upon the surviving original examples.

One of the most important decorative elements, however, was missing. The Royal Achievement of Arms that was the focal point of the stern of all Restoration warships of this period was absent. Invariably this would consist of an impressive carving of the coat of arms of the Stuarts, flanked by the lion and unicorn, surrounded by flowing acanthus leaves, and bearing the motto of the Order of the Garter: *honi soit qui mal y pense* inscribed in miniature letters on a scrolling banner. There was, therefore, no mystery about what this carving should look like, but making a convincing replica would be a great challenge, even to Philip Wride, who is a master at seventeenth-century style carving and to whom we entrusted all the other carved decorations.

Our problem was solved in a way that was so unexpected and remarkable that it hardly seemed possible. A seventeenth-century boxwood carving of the Royal Arms of exactly the right size and date, and reputed to have come off the stern of a ship model, came up for sale in a Sotheby's furniture auction in London at just the moment when Philip was to begin carving his own replacement. The sale consisted of the personal collection of the late London dealer Ronald Lee. We had known Mr Lee, who ran an extraordinary establishment at Bruton Place, for many years. Nearly all the items in his relatively small shop were of exceptional quality and rarity, and many found homes in prominent museums. Mr Lee had once sold an Admiralty Board ship model (for a more complete account, see Chapter 9), and we always made a point of visiting when we were in London. On one occasion, we were shown a small boxwood carving of the Royal Arms in a small wooden frame, which Mr Lee said had probably come from the stern of a seventeenth-century ship model. It was not for sale, however, being part of his personal collection. Years passed, and on one visit we found the premises empty. On making enquiries, we were saddened to learn that Mr Lee had passed away. We were therefore intrigued when, a year or so later, we noticed an upcoming sale of items from his personal collection. In the sale catalogue, lot 133A was the little

HISTORIC SHIP MODELS

The equestrian figurehead on this model is a very fine example of a form that was common among Carolean 1st rates, with similar figureheads fitted on the *Prince*, *Britannia* and *Royal Sovereign*. In this instance, the rider bears an intriguing resemblance to King Charles II. The knee of the head is fitted with an unusually decorated accommodation for the fore tack leads, in the form of an upper torso with outstretched arms. The trailboard is carved with a whimsical scene of a hound (obscured by the gammoning ropes) in pursuit of a hare.

On English ships, the belfry is always placed in the foredeck bulkhead. The impressive belfry canopy is original and is decorated with gilded dolphins alternating with cornucopia, presumably an allegorical reference to the bounty of the sea. The jeer capstan in the waist is pierced for four through bars and has a spindle seated in the deck below. The fish davit passes across the foredeck and is shackled to the starboard sheet anchor.

boxwood carving of the Royal Arms, and it came up just as we were deciding how to deal with the missing stern carvings on the *Royal James*. We called the auction house to obtain accurate overall measurements and were delighted when they turned out be a perfect match for the missing carving.

Encouraged by the remarkably low auction estimate, we quickly resolved to buy the little carving. We expected that no one would likely want this object more than we, so we were confident that we would outbid the competition. We set what we considered to be a very generous bid limit, many times above the upper estimate. When the carving came up, Arnold was bidding on the phone. The auction house agent assisting Arnold remarked that it was a lovely little piece and that there had been a lot of interest. Readers familiar with auction house jargon will know that this phrase usually means an item will be hotly contested and almost always exceed its high estimate, sometimes by orders of magnitude. This was our first hint that our bid might be in trouble. A minute later lot 132 came up and sold well within its estimate. Then our lot came up. Arnold entered the bidding during a pause. Within seconds he was bidding against someone in the room and up it went to near our predetermined limit. There was another pause, but then a new bidder entered and Arnold found himself authorising bid after bid beyond our limit. Another pause, and then another bidder entered. Arnold became anxious about what was rapidly becoming a very extravagant carving. But at an auction one can always take comfort, whether rightly or wrongly, that even when one has paid too much for something, there was at least one other person at the time willing to pay nearly as much. Arnold also knew that we had never regretted any model we ever bought, just the ones that we had missed, and where would we ever find a more perfect carving than this? When the dust settled, we had the highest bid, though we were stunned at how high it had gone and how quickly. Sotheby's agent then congratulated Arnold. Again, those familiar with auction parlance will recognise this as a nearly infallible sign that in Sotheby's estimation, we had paid too much and they were glad of it. However, we had succeeded in acquiring a superb, authentic seventeenth-century carving, possibly originally made for a ship model, which we could now use to adorn the stern of the *Royal James* model.

In restoring the *Royal James*, we decided to have it fully rigged. Although we do not know whether the model had been rigged originally, we were interested in taking advantage of Philip's enormous modelling skill and in creating the most accurate rigging that had yet been put onto a Restoration model. Also, rigging would not alter the hull in any way and could always be removed by future generations if so desired. What made this unique project so appealing was our acquisition of a seventeenth-century manuscript book by Edward Battine, entitled, *The Method of Building, Rigging, Apparelling and furnishing his Majesties Shipps of Warr according to their Rates ....* in which the measurements of all the rigging ropes and blocks as well as dimensions of masts and spars are given for the *Royal James* of 1675. Battine was

Surveyor at Portsmouth dockyard, and he took great pride in preparing a manuscript providing extremely specific details concerning the building and rigging of late seventeenth-century warships. At his own expense, he prepared several copies of his little manuscript volume. The manuscripts, nearly all identical, are illustrated with numerous highly detailed tables based on measurements taken from representative ships of the period. The rigging particulars for the 1st-rate ship in these tables are taken from the *Royal James* of 1675. Battine's tables, however, have not to our knowledge been published. Battine had the multiple copies of his manuscript sumptuously bound and given as gifts to influential Navy officers, including James II (then Lord High Admiral), Samuel Pepys, King Charles II, and at least two copies to George Legge (Lord Dartmouth). Pepys was possibly jealous of Battine's work, because he was in the process of preparing a similar, though less extensive, volume of his own and had been scooped. Pepys wrote Battine a scathing letter damning him with faint praise and rebuking him for sending a document to the King that was full of so many errors.[7] If only he had consulted with Pepys first, Pepys disingenuously remarks, Battine might have saved himself embarrassment.

In another example of remarkable coincidence, one of the eight known copies of Battine's manuscript came up for sale as we were considering the rigging of the *Royal James*. It was one of George Legge, 1st Baron of Dartmouth's copies dated 1685, and bore his name in the dedication. Dartmouth was Master of the Horse, Governor of the Tower of London, Admiral and Commander-in-Chief of the Fleet to James II, and a highly influential naval personage. It was in wonderful condition, within its original beautiful roan leather binding covered with gilt devices made by the 'Naval binder', active between 1673 and 1689,[8] and still secured by its original silver-gilt clasps and catches. We were able to buy this copy, photograph the relevant tables, and send them to Philip Wride, who used Battine's measurements, together with Deane's own as listed in his *Doctrine of Naval Architecture*[9] in the masting and rigging of the model. The rigging is in silk, as this is the material that was used in the seventeenth century to rig models. Silk gives the rigging a fine clean surface and a very convincing miniature rope appearance. The lines have all been spun with the correct twists on a miniature ropewalk and have correct scale dimensions. A total of sixteen different diameter ropes were made and over 5 miles of silk thread were used to spin all of the miniature lines.

CONSTRUCTION

Scale: 1/48 Hull length: 52in

This model is one of only two surviving examples of Restoration 1st-rate ships. The other one is the famed *Prince* model in the London Science Museum. These are interesting models to compare. The well-known model of the *Prince* is a beautiful and well-preserved model that matches nearly perfectly drawings and paintings of the ship herself and is therefore firmly identified. This model was acquired by the Science Museum in the late nineteenth century at a provincial auction for £10.[10] Interestingly, the model was originally unrigged, and the masts and rigging were added in the museum workshops. The models of both the Prince and *Royal James* have similar finely carved equestrian figureheads. The Royal James is the larger of the two, and has a remarkably long rake at the bow, a relatively old-fashioned feature even for 1671. The model of the *Royal James* is the only known Restoration 1st rate remaining in private hands, and it is unlikely that another, if it exists, will ever be permitted to leave England.

Construction on the *Royal James* began at the Portsmouth dockyard in 1668, and the model may date from about this time. The equestrian figurehead is original and is a superb example of restoration carving, and while the namesake of the ship is James I, it bears a striking resemblance to Charles II. It is likely that Deane thought it wiser to flatter a reigning King than to honour one long dead. The canopy over the belfry is a particularly fine and unusual example consisting of carved supporters in the form of overflowing cornucopias alternating with dolphins, allegorically suggesting the bounty of the sea. The knee of the head features a unique carving that decorates the bored holes through which the fore tack leads are reeved. This gilded decoration takes the form of the upper torso of a man with arms outstretched to embrace both leads. There is no similar example on any dockyard model, and the only instance where we have seen decorations on the knee of the head of an English ship is in Deane's manuscript *Doctrine of Naval Architecture* written in 1670. The illustrations in this treatise further support the conclusion that this model is his work.

Right: The decorated portside entry is flanked by caryatids supporting a decorated canopy. This was the primary access route, and although there would have been a handhold rope for assistance, negotiating the steps in rough seas would have presented a significant challenge for a non-sailor.

The model is fitted with diagonal underlaid catheads, a very unusual feature on a seventeenth-century three-decker. The upper surfaces of the cat-tails have been notched where they pass under the deck beams to allow the cat-tails to lie flush with the underside of the fore deck. There are entry ports on both starboard and port, rather than on the port side only. This is a feature known to have appeared in the 1670s, and to our knowledge this is the earliest example. The painted quarter gallery decorations include an interesting pair of caricature faces on both starboard and port that appear to be old hags. The carved brackets at the cove of the stern take the form of the usual crouching torsos, but they have the unusual feature of having lion masks at their feet as well as their heads. All other examples have them only at their heads. There are two original capstans, the jeer capstan is at the waist and the main capstan is below on the main deck just abaft the mainmast. Both capstans are pierced for four through bars at staggered levels, one of the last surviving examples of an old form of capstan that was replaced by the drumhead type around 1680. The forward portion of the poop deck does not extend to the ship's side. This deck is accessed by a pair of low narrow gangways passing over the quarterdeck guns from the bulkhead to a point almost halfway to the stern. This allowed adequate headroom for the officers' cabins under the poop while keeping the sides of the hull as low as possible. Frank Fox, an authority on the Restoration battle fleet, called our attention to a similar feature on the triple-wale model at the NMM and a van de Velde drawing suggesting that it may also have been present on another of Deane's 1st rates, the *Royal Charles* of 1673.

Left: The masses of carved and gilded decoration that adorned the sterns of Carolean ships were most evident in large three-deckers such as this. Life at sea was every bit as socially stratified as on land, and given that the most glorious decorations enclosed the stern, it comes as no surprise that this is where the flag officers were quartered.

Literature

The following references include photographs and descriptions of this model:

Culver, Henry, *Contemporary Scale Models of Vessels of the Seventeenth Century* (New York: Payson & Clarke Ltd, 1926), p. 13.

Fox, Frank, *Great Ships the Battlefleet of King Charles II* (London: Conway Maritime Press, 1980), p. 114, plates 129 and 130.

Leetham, Arthur, *Official Catalogue of the United Service Museum* (Southwark: J J Keliher & Co, 4th edition, 1914), Additions to the catalogue, p. 36 (as the *Albemarle*).

Nance, R Morton, *Sailing-Ship Models* (London and New York: Halton & Truscott Smith Ltd, 1924), p. 70, plate 35 (as the *Albemarle*).

Nance, R Morton, *Sailing-Ship Models* (London and New York: Halton & Truscott Smith Ltd, 1949), p. 55, plate 32.

Philbin, Tobias, & Endsor, Richard, *Warships for the King* (Florence, OR: SeaWatch Books LLC, 2012), p. 110.

Robinson, Gregory and Anderson, R C, 'Identification of Men-of-War VI', *Mariner's Mirror*, 2 (1912): pp. 340–341, plates 17–18.

Winfield, Rif, *British Warships in the Age of Sail 1603–1714* (Barnsley: Seaforth Publishing, 2009), p. 11.

Winfield, Rif, *First Rate* (Barnsley, Seaforth Publishing, 2010), pp. 29, 34–35, 66, 86, 134.

Exhibitions

London, Royal United Service Museum, Before 1908–1948.

'Ship Models form the Great Age of Sail 1600–1850', Greenwich, England, National Maritime Museum, 18–20 April 1996.

Washington DC, National Gallery of Art, *Water, Wind, and Waves: Marine Paintings from the Dutch Golden Age*, 1 July–25 November 2018.

Salem, Massachusetts, Peabody Essex Museum, September 2019–present.

Historical Perspective

SIR ANTHONY DEANE, SHIPWRIGHT AND MODEL BUILDER

This model is the design for the 100-gun 1st-rate Restoration warship the *Royal James*, made by Anthony Deane. Deane built two ships named *Royal James*, the first one was completed in 1671 and had the misfortune to be burned and sunk by the Dutch at the Battle of Solebay within one year of its launch. Deane was immediately commissioned by the King to build a replacement ship. In October 1673, before the completion of this second *Royal James*, the Admiralty ordered that the distance between the guns be increased and the total armament reduced in all 1st rates. These changes delayed construction, and the *Royal James* was not completed until 1675. Charles II himself attended the launching ceremony as a tribute to Deane. Charles II had been sufficiently impressed with Deane's achievement in building the previous *Royal James*, as well as the *Royal Charles*, that he had knighted him for his service to the Crown, and he became the only English shipwright ever to attain this honour. Our model most likely represents the *Royal James* of 1671. It closely resembles surviving van de Velde depictions of this ship and has several features that, while appropriate for a ship of 1671, would have been anachronistic for a ship of 1675. There are some differences in gun port arrangement and decoration between the model and the completed ship as shown by van de Velde. These are accounted for by typical alterations during construction; similar differences exist between the model of the *Prince* of 1670 and the final version.

Anthony Deane and Samuel Pepys became good friends, and this friendship helped propel Deane to the forefront of English shipwrights and to the attention of the King. The lifelong friendship began when Deane, who was then an Assistant Shipwright at Woolwich dockyard, gave Pepys, who was the influential Clerk of the Acts and a member of the Navy Board, the gift of a ship model. Pepys records his first meeting with Deane in his Diary entry for 11 August 1662:

> Mr Deane, the assistant at Woolwich came to me, who I find will discover to me the whole abuse which his majesty suffers in the measuring of timber, of which I shall be glad. He promises me also a modell of a ship which will please me exceedingly, for I do want one of my owne.[11]

Then, as now, gifts to men of influence can help advance one's career. Deane was quick to satisfy Pepys' request, as recorded in his diary entry one month later on 29 September 1662:

> Went home, where I find that Mr Deane of Woolwich hath sent me the modell he had promised me. But it so exceeds my expectations that I am sorry almost, he should make such a present to no greater person; but I am exceedingly glad of it, and shall study to do him a favour for it.[12]

Interestingly, Deane had not built a warship for the Navy at the time he gave his model to Pepys. Pepys and Deane became good friends, and on Pepys' recommendation, Deane was promoted to be master shipwright at Harwich in 1664, when he was only twenty-six. Later that year he began work on his first warship, the *Rupert*, a large two-decker. When Harwich was closed at the end of the Second Anglo-Dutch War in 1668, Deane was shifted to Portsmouth, and it was here that he built his first 1st rate, the *Royal James* of 1671.

Deane's pre-eminence as a naval architect owed much to the influence of Pepys, but would not have been achieved if it were not for his own ingenuity and dedication. He was one of the first to experiment with lead sheathing to protect against wood-boring teredo worms. His pioneering experiment in the *Phoenix* of 1671 failed, however, because of the electrolytic action induced by the iron nail fasteners. He was also the first to use iron knees, which were incorporated in the hull of the *Royal James* of 1671, a measure designed to economise on compass timber (the naturally curved tree limbs that were always in short supply). This innovation also failed to take hold, possibly because the *Royal James* burned and sank at the Battle of Solebay soon after it was launched, too soon to fully evaluate its performance. Pepys also credits Deane with inventing a system for calculating the displacement of a ship prior to building. Deane became a member of the Royal Society, of which Pepys was not only a member, but a one-time president (Sir Isaac Newton's *Principia* was published during Pepys' tenure as president and bears a dedication to him). It was at Pepys' insistence that Deane wrote his *Doctrine of Naval Architecture*, illustrated with his own beautiful drawings and plans. The manuscript for the treatise remains to this day in Pepys' library now housed at Magdalene College, Cambridge, but it was never published in Deane's or Pepys' lifetime. The treatise was eventually published in 1977, nearly 300 years after it was written and long after it could have had any impact on shipbuilding practice. Deane's friendship with Pepys outlived both their careers, and they remained close into old age, enjoying in Deane's words, 'The old soldier's request, a little space between business and the grave.'

LUCIFER AT THE HELM: TERROR TACTICS IN THE AGE OF SAIL

Seasoned oak, pitch, canvas and rope, the raw materials for constructing a wooden warship, are also ideal ingredients for building a bonfire. This coincidence was not lost on the strategists of the seventeenth century, who routinely employed floating incendiary weapons known as fireships. These demonic devices were either small warships or merchantmen specially modified to burn fiercely. In battle they would be filled with combustibles and sailed toward key enemy ships by an intrepid volunteer crew. The idea was to ignite the fuses at just the right moment to allow the crew to escape in an open boat while the unmanned fireship entangled its target. Substantial bounties were offered to induce sailors to volunteer for such extremely hazardous work. Of course, the enemy would do everything it could to repel such attacks, including firing on the approaching vessel, pushing it way with long poles, and trying to sink or cut loose the boat intended for transporting the fireships' crew to safety. Much of the impact of fireships derived from the psychological effect they had on sailors, who could hardly imagine a more terrifying fate than to die in a conflagration. Because of the difficulties of successfully attacking a flagship with fireships, only two were destroyed this way in the entire Third Anglo-Dutch War, and the most memorable and infamous was the burning of the *Royal James* at the Battle of Solebay.

Ornate as the quarter galleries are, they served as latrines for the officers and commanders. The circular openings at the lower and maindeck levels were presumably apertures for small arms fire, though there is at least one seventeenth-century painting that shows a cannon barrel protruding from such a port. The lower canopy is adorned with a carving of two dolphins with entwined tails.

The Battle of Solebay, fought on 28 May 1672, was the first and largest engagement of the Third Anglo-Dutch War, and was the first and last battle for the *Royal James*. On that day the *Royal James* was the flagship of Edward Montagu, the Earl of Sandwich, whose behaviour was largely shaped by events that transpired the day before. The Duke of York had brought the combined fleet of

Just abaft the mainmast, centred in the bulkhead of the coach, can be seen a projecting portico with double doors that provides an entrance to a descending staircase to the middle gun deck. On the outside bulwarks, there are curved hancing pieces that mark points of transition in the height of the ship's sides. Other ships of this period had hancing pieces carved of vertical timbers instead. At the very stern, set against the taffrail, are the trumpeters' cabins.

England and France to Sole Bay, where they loaded provisions for a planned foray off the Dogger Bank. On 27 May, in the face of a freshening on-shore breeze, Sandwich cautioned the Duke of York that the fleet should make for the open sea lest the Dutch attack them in their exposed position. The Duke rebuked Sandwich in the company of the commanders of the fleet. The exact wording is lost, but we know that the Duke of York wittily denounced Sandwich's apparent faint of heart, and the Duke of Albemarle who was in attendance went so far as to accuse Sandwich of cowardice. The effect on Sandwich was dramatic, as one can imagine in the face of accusations coming from such quarters, namely from the Lord High Admiral who was also heir to the throne. Moreover, this incident came within weeks of a Council of War at which Sandwich had urged caution and argued against risking a fight with the Dutch while the English fleet lay close to the Goodwin Sands. Sound advice, no doubt, but not what the assembled bullish fleet commanders had in mind. When Sandwich's courage was questioned again on the evening of 27 May, his usually cheerful demeanour changed abruptly, and according to a witness, he became grave and gloomy. When the Dutch fleet was sighted off Sole Bay early the next morning one can only imagine the brave determination with which Sandwich prepared for his part in the ensuing battle. In action, a ship embodies its commander, and the *Royal James* was nothing if not bold and decisive. The *James* was made to charge directly at the enemy, to repel attack after attack, to maintain a tireless cannonade, and to fight beyond endurance. In the end she slipped beneath the waves and into the annals of naval history with honour secured.

On the morning of 28 May, the Dutch approached slowly due to the light air, which gave the allied fleet time enough to cut cables and prepare for battle. The Blue Squadron was closest to the Dutch fleet, which was headed north, so the Allied fleet also headed north and prepared to fight in reverse order, with the Blue Squadron under Sandwich taking the lead instead of bringing up the rear as was customary for the Blue Squadron, while the Red Squadron, under the Duke of York took up the middle, and the White Squadron, under the French Admiral, Comte d'Estrées, actually headed south, away from the rest of the fleet. This may have been because the White Squadron was technically supposed to be in the van, and the French commander may have expected that the British squadrons would follow his lead, but history has passed a less charitable verdict, and the Comte d'Estrées is usually remembered as a villain who cowardly deserted the fight.

The Battle of Solebay was the fiercest battle in all of the three Anglo-Dutch Wars. Two hundred and ninety-four ships bearing over 50,000 men fought on that beautiful spring day with light offshore breezes and an ocean 'as calm as a milk-bowl'. The fighting was so intense that the drumming beat of cannon fire could be heard miles inland from seven o'clock in the morning until around six o'clock at night, and the coast was blanketed with clouds of smoke and the acrid smell of gunpowder. The battle began at six in the morning, when van Ghent, Admiral aboard the *Dolfijn*, bore down upon the *Royal James* and delivered an opening broadside. Sandwich returned fire and the ships battered each other for over an hour. This is when the first fireship was sent toward the *Royal James*, but a well-aimed broadside from the *James* sank her. A second fireship was dispatched, but her sails and rigging were shot away and she missed her target. By now a dense pall of smoke surrounded the *James*, and only the flag at the main top was visible to observers nearby. Several other men-of-war of van Ghent's squadron joined the engagement. One was so riddled with shot from the *James*'s guns that she eventually sank. The Dutch Captain van Brakel, who had helped the Dutch humiliate the British in the raid on Chatham in the Second Anglo-Dutch War, broke away from van Ghent's squadron and, ignoring signals to re-join, bore down on the *Royal James* instead. So impetuous was van Brakel that his ship, the *Groot Hollandia*, actually struck the *Royal James* near the bow, and the two ships became entangled by intertwined spars and rigging. Thus began over one hour of near continual close fighting. It has been estimated that within the first few minutes 300 men were killed or disabled on the *Royal James*. It is difficult to imagine the carnage with limbs, heads, and body parts crushed, severed, or simply carried away by massive hurtling cannon balls – or penetrated by deadly wooden splinters – or crushed by careening cannon or falling spars. A cannon ball from the *James* struck and killed van Ghent in the *Dolfijn*, but the ship fought on. After one and a half hours, the *Groot Hollandia* became disentangled from the *Royal James*, and they drifted apart. Three sailors from the *Royal James* who had climbed aloft to cut the tangled rigging crossed over to the *Hollandia* and took down her pendant just before the ships parted, but they did not reboard the *James* in time and became prisoners of the Dutch. Van Brakel admired their audacity and is said to have rewarded them with 100 ducats each to inspire his own men. The *Hollandia*, however, had been reduced to a wreck. Van Brakel was injured and most of his officers and crew were dead or wounded.

At this point in the battle, the *Royal James* was engaged with several Dutch vessels and enveloped in clouds of cannon smoke that hung in the gentle breeze. This, and the preoccupation of what remained of the crew who were firing guns below decks, may have helped yet another Dutch fireship approach unseen. Some accounts claim four fireships were sent against the *James* that day, others claim three, but there is no doubt that only one actually reached her. By the time the danger was realised, there was nothing that could be done. The *Royal James* was a battered hulk, her crew reduced to a fraction of her full complement, and the wind too light for manoeuvres. The fireship drifted alongside and the flames licked the rigging and ignited the sails. The *James* blazed brightly from two to four o'clock in the afternoon and continued to burn for two hours more until she slipped below the waves. Her captain, Richard Haddock, climbed out of a porthole and swam for 2 miles until he was rescued. Accounts of Lord Sandwich's final minutes differ. Some witnesses say that Sandwich stayed aboard his ship to the end, others that he cast off in a barge to shift his flag to another ship, but the barge, being overloaded, capsized and he drowned. His body was recovered later, and his final resting place was not beneath the waves, as he may have expected, but in London, where he was buried with honours. Among the many tributes inspired by the life and death of Edward Montagu, the First Earl of Sandwich, is a poem by John Campbell, author of *Lives of the British Admirals* that ends thus:

> *Go, serve thy country, while GOD spares thee breath;*
> *Live as I liv'd, and so deserve my death.*[13]

It is a remarkable coincidence that of all the British three-deckers that fought that day, dockyard models survive of three of them: the *Prince* in the Science Museum, London, the *Royal James*, in our collection, and the *Saint Michael* at the NMM, Greenwich. Another great ship that fought that day, the *Royal Charles*, survived in the form of a model that lasted 275 years, before succumbing to a direct hit from a German bomber during the Battle of Britain in the Second World War.[14] However, a fine and accurate reconstruction of this model now takes its place at Trinity House, London. Though the ships are all long gone, what a remarkable commemoration could be staged by exhibiting these magnificent models together. There would be no better way to invoke the spirit and grandeur of one of the most memorable events of the age of sail.

The *Royal James* of 1671 achieved everlasting fame through the dramatic circumstances of its demise. The *Royal James* of 1675, although begun during the Third Anglo-Dutch War, was completed after peace had been declared. She participated in no great sea battles and achieved no real distinction as the *Royal James*. Following the Glorious Revolution in 1690 that saw the Catholic James II exiled to France and the Protestants, William and Mary, jointly assume the throne, the name James became a definite liability for a royal ship. The *Royal James* was renamed *Victory* in 1691. She was not the first British ship to bear this name, but since there was no ship named *Victory* at the time, it was available and suitable. In 1714 she was briefly named *Royal George* before reverting to *Victory* again in 1715. It was as the *Victory* that she was finally broken up in 1721, having enjoyed a remarkably long life for a wooden ship. In 1737 a new *Victory* was built, and in 1744 she sank, taking her captain, Lord Balchen, with her to the bottom (see Chapter 31). Not to be discouraged, the Admiralty ordered a new *Victory* to be built, and she was launched at Portsmouth in 1765. This incarnation of the ship would take her captain, Lord Horatio Nelson, into the Battle of Trafalgar and secure for her and him, everlasting glory as, respectively, Britain's greatest warship and naval hero. The *Royal James* of 1671 is thus part of the lineage of arguably the most famous ship in the history of the British Navy.

References

Callender, Geoffrey, *Sea Kings of Britain* (London: Longmans, Green and Co, 1939), pp. 39–46.

Campbell, John, *Lives of the British Admirals*, Vol. 2 (London: Printed for C J Barrington, 1812).

Colliber, Samuel, *Columna Rostrata* (London: Printed for R Robinson, 1727), pp. 215–25.

Fox, Frank, *Great Ships the Battlefleet of King Charles II* (London: Conway Maritime Press, 1980), p. 110.

Fox, Frank, *A Distant Storm* (Rotherfield, England: Press of Sail Publications, 1996).

Right: The remarkable height of the stern of restoration warships can be appreciated from this perspective. The admiral strutting his quarterdeck or poop would have enjoyed a panoramic vista, provided he was immune to vertigo.

THE *ROYAL JAMES*, 1ST RATE OF 1671

Instead of the usual boxwood on the topsides, this model features walnut. This reflects a fashion introduced by joiners and cabinetmakers of this period, who began incorporating walnut veneers in their best case furniture. The topsides are unscribed with no markings to indicate planks or joins. The gun port arrangement is unusual in having none in the waist. Several ships of this period were originally built this way but were subsequently pierced for a full complement of guns. The shape of the wreaths around the ports, with their square inner edges, is a distinctive feature.

CHAPTER 2

A Charles II 5th rate c1680

Acquisition

IT IS RARE TO find a seventeenth-century ship model in its original display case, but it is rarer still to find an empty display case and years later, the model it contained. Much like an orphan reunited with a long-lost parent, we can point with pride to the combination of determination, singleness of purpose and good fortune that ultimately led us to one of our proudest collecting moments. In 1968 the Fairfax family of Acomb, Yorkshire, sold this model in its original case to Sussex antique dealer Peter S Westbury. It was subsequently sold to J M Williams, a Massachusetts antique dealer, in May of that same year. We first learned of this model when an illustrated advertisement appeared in *Antiques* magazine in 1978. Henry was an ophthalmology resident in Boston at the time, and the ad, placed by Mrs Williams for a seventeenth-century lacquer and gesso glazed cabinet, caught his attention. The accompanying photo showed a seventeenth-century Admiralty Board model in the cabinet, but the ad unfortunately, said that the model had already been sold. Henry immediately drove to Beverly, Massachusetts, to see it and heard the sad tale of how the model and case had become separated. When the late Mr Williams bought the ensemble, he was interested in the remarkable cabinet and had little regard for its extraordinary contents. He 'tore' the cradles out of the case and sold the model to a marine antiques dealer shortly after he acquired it. Ten years later, Mr Williams died, and Mrs Williams couldn't remember to whom her husband had sold the little ship. We bought the empty case and began to look for the missing model.

Advertising for the model, which we did using an old photograph, did no good. After about a year, we were delighted to get a call from Mrs Williams, who had come across an invoice with the identity of the buyer. It had been sold to marine antiques dealer Karl F Wede of Saugerties, New York, back on 5 May 1969. This was not altogether good news for us since we knew that the remaining contents of Karl Wede's shop had been sold at auction several years before, and neither Mr Wede nor his business were still around. We quickly confirmed that he was no longer listed in the phone book.

We had actually met Mr Wede on one occasion when we were travelling to upstate New York with our father. We were driving along the Hudson River and had stopped at his shop. Henry remembered that Mr Wede was a native of Germany and that he asked our father an odd question – he wondered if our father would be interested in buying the business so that Mr Wede could move

This handsome ship model display cabinet dating from around 1680 is the oldest and most ornate surviving example known. The exterior is japanned in black with gilt floral and geometric designs, and the decorative carved mouldings, as well as the cradles supporting the model, are silver gilt coated with gamboge-tinted shellac to produce a gold effect. The interior of the cabinet is painted in Venetian red.

Here we fitted launching flags to the model to recreate the original appearance.

back to Germany. He evidently never found a buyer, but Henry wondered if he might nevertheless have fulfilled his wish to emigrate. We called the Post Office in Saugerties and spoke to the man who delivered mail along the route RFD #3. He remembered Karl Wede. Furthermore, he confirmed that he hadn't died but had moved away. Having come so close to tracking him down, we were sorely disappointed when we learned that he had left no forwarding address.

We were temporarily stymied, but eventually thought of calling the Mystic Seaport Museum and inquiring about Mr Wede. A friendly staffer consulted a list of some sort, and we were told that they *did* have an address for a Karl Wede, but 'unfortunately it was in Germany!' After two minutes and a quick call to German directory information, Henry was talking to Mr Wede himself!

However, our search for this model was not to be concluded so easily. Mr Wede remembered the model, and he recalled selling it, but he regarded the sale to be a confidential matter and would not tell us anything about the buyer. He agreed to write a letter to the mysterious buyer, including one we sent along, but we never had a response. We waited one year and called Mr Wede again. We pleaded that we had the original display case and several parts of the model, including gun port lids, but still to no avail. We continued to call annually for several years until we finally got some positive news. Mr Wede said that we were at last going to be able to see the model, as the owner had died and his collection of models and firearms was going to the Smithsonian Institution. This was all we needed to know. The Smithsonian has an English Navy Board model in its collections, but we knew it had never even been on exhibit, so we doubted that they would have much interest in another English ship model. We promptly called the Institution. We spoke to someone who knew about recent bequests, and she did, in fact, know of a collection of arms and ship models that they had been offered. Much to our delight, she said that they accepted the firearms but had declined the models. Their legal department then let the donor know of our interest, and we soon received a call from a woman outside Philadelphia. She had arranged for Christie's to pick up the collection of models her late husband had assembled, and yes, there was one that fit our description. We offered her substantially more than Christie's had

The tall stern is topped by a pierced taffrail. Prominently carved in the centre of the taffrail is a Charles II monogram, flanked by cherubs riding upon dolphins. Between the quarterdeck lights there is a royal Stuart coat of arms surmounted by a crown. One would expect the Royal Arms to be flanked by a lion and unicorn, but instead there is an unusual shield-like surround with cherub heads. Just below the upper stern lights, there are a pair of carved faces identified as sailors by their hats. Tudor roses, gargoyle masks, caryatids and foliate panels and terms complete the stern carvings.

thought it would bring at auction, and she agreed to set it aside when the other models were picked up in two days. That weekend Henry drove to Philadelphia, and there was the missing model! It had never been put in another case, and many of the carvings had fallen off, but they had all been collected and put into a cigar box! Later that day, our offer accepted, we proudly reunited the model with its original case.

Provenance

Rear Admiral Robert Fairfax (1666–1725) most likely acquired this model in the latter part of the seventeenth century. It remained in his family estate in Acomb, Yorkshire, until it was sold in 1968 and exported to America.

A crowned Tudor rose, the heraldic emblem of England, is featured in the central quarter gallery carving just below the row of lights. The upper finishing is capped by a carved urn containing a bouquet of roses. The winged cherub was a commonly used device for the lower finishing of the quarter gallery. In this case, a sprig of flowers appears below the cherub, and there is an acanthus leaf decoration to the upper stool, giving the quarter gallery a distinctly sylvan theme.

Description

CONDITION

This is one of only a handful of models known that represent small two-deckers from the reign of Charles II. The outstanding condition of this model is especially remarkable given that it has survived three and a half centuries. It is essentially complete and original in all details, including the unique winged cherub-head giltwood cradles. Fragile objects such as this do not survive well without protection, and the preservation of this model results from it having remained in its original case for over 300 years. The beautiful japanned, giltwood, polychromed, and glazed display cabinet is itself a remarkable survival, and the oldest ship model display case yet discovered. It was built with sufficient height to allow mounting of flagstaffs at each mast hole, which would have carried launching flags. Unfortunately, these have not survived.

As related above, the model was removed from its cabinet for a span of about seventeen years. During this period of separation, the animal hide glue that held some of the carved decoration in place began to fail, and the model literally began falling to pieces. Luckily, all the loose parts were saved, so that after we acquired the model, Philip Wride (a model shipwright of extraordinary ability) was able to reattach the parts and thus restore the model to its original 1660s appearance.

CONSTRUCTION

Scale: 1/48 Hull length: 32in

This model is one of a small number of interesting 5th rates built during the reign of Charles II that had an incomplete tier of upper deck guns and with none in the waist (a so-called 'gunless-waisted' ship). This feature was common in merchantmen of this period, as well as in 5th-rate and small 4th-rate warships.

The topsides are done in walnut rather than the more usual boxwood, a construction feature shared by several other models of this early period. Additional early features include the long rake of the head knee, the square inner sides of the port wreaths and the absence of panels on the bulkheads.

The model is fully armed, with the usual seventeenth-century-style guns. These consist of turned wooden gun barrels mounted on simple stepped wooden carriages, carved out of one piece, and lacking trucks. Differences in the height of the decks at different gun port locations meant that each gun had to be carved to fit its particular site. Some of the carriages are therefore quite shallow, while others are deep. (Apparently this was also the case for some seventeenth-century ships themselves, and gun carriages were often numbered and constructed specifically to fit into particular gun ports.) The taffrail displays the Charles II cipher, crowned, framed and flanked by dolphin-riding cherubs. The stern badge contains the King's arms, and is flanked by a pair of cherub heads below, with neither the lion nor unicorn in evidence. Two stern lights pierce the roundhouse, with medieval-looking, long-haired, hat-wearing male heads carved below. Where there are no stern lights, carved floral panels decorate the spaces between the vertical timbers and slender caryatids with lion masks adorn the curved counter timbers. Two stern chase ports are open with menacing cannon protruding. Each quarter gallery is centred on a large arched light above a carved panel decorated with the crowned rose of England. There are winged cherubs below the counters resembling the winged cherubs that form the cradles that support the model in its case. Both quarter galleries are surmounted by a carved vase of Tudor roses. The model is equipped with a drumhead jeer capstan, which is among the earliest examples of this fitting on a model.

Nearly all English warships smaller than 2nd rates had lion figureheads, and this is no exception, but the lion has a little female companion to fill the space between the figurehead and the carved bracket on the first head timber. The beak platform and adjacent head grating curve upward to meet the main headrails, producing a complex curve that must have been a challenge to construct. The knightheads are simple, undecorated timbers. The hancing pieces are carved in the form of crouching lions.

~ Literature ~

The following references include photographs and descriptions of this model:

Davies, J D, *Pepys's Navy, Ships, Men & Warfare 1649–1689* (Barnsley: Seaforth Publishing, 2008), p. 56.

Franklin, John, *Navy Board Ship Models 1650–1750* (London: Conway Maritime Press Ltd, 1989), pp. 14, 61, 85–88.

Kriegstein, Arnold and Henry, 'The Kriegstein Collection of British Navy Board Ship Models', *Nautical Research Journal*, 38, 4 (1993), p. 218 and plate 1.

Winfield, Rif, *British Warships in the Age of Sail 1603–1714* (Barnsley: Seaforth Publishing, 2009), p. 161.

The lion figurehead was standard on all but the largest ships, but in some, as in this example, he has a small female companion. The seats of ease for the convenience of the sailors can be seen on the beakhead, as can the unusually curved beak platform and head grating.

Historical Perspective

The unique and ornate period display cabinet built for this model conforms remarkably well to the first specifications ever recorded for a ship model display case. Samuel Pepys had his joiner build a set of glazed bookcases for his library, which are the first such cabinets ever constructed, and he adopted the same approach for the protection of his models. On 2 May 1677, Pepys wrote the following letter to Phineas Pett, master shipwright at the Chatham dockyard:

> I take this occasion of giving you and your Lady thanks for your great civilities to me [at Chatham] and more particularly for that of the present you are providing for me and which I shall labour to put a due value upon; and as one instance of it, do adventure to trouble you a second time about that which I took the liberty of observing to you when I was with you, namely, the thickness and breadth of the stile and rail where the glass is to be set, which I do by all means desire may be made as small as may be; those which I have for my book-presses [cases] here not being above ¾ of an inch in thickness either way, struck on the outside with a small astrical or half-round.
>
> True it is, the glass by that means will become the larger, but I shall pray you to leave the providing of that to me, and suffer the sashes to be brought up hither empty, putting you also in mind of having one side made up without glass, to be laid pure white within-side, with a moulding only round it to be gilt as the rest is; without which, or some such ornaments, I fear it may appear somewhat too plain.
>
> Let me also entreat you that the pedestal, or whatever it is upon which the model is to stand, may be moveable, that upon occasion it may be taken out of its cabinet and set upon a table for the better looking round it … I purposing to make more use of it than barely for the entertaining of my eyes, and consequently to be indebted to you for somewhat more than a piece of furniture, though that in itself were very valuable.[1]

We are not certain when this model became part of the Fairfax legacy, but it most likely was acquired by Admiral Robert Fairfax. The Fairfax family rose to prominence as the result of the exploits of the admiral's grandfather, Sir William, in the civil war that preceded the Commonwealth period. The monarchy had been renounced in favour of parliamentary rule, and war had broken out between Loyalists and Parliamentarians. Sir William Fairfax was a staunch Parliamentarian and died a valiant death while raising the siege of Montgomery Castle on 18 September 1644. His wife responded to news of his demise with the statement that 'she grieved not that he died in the cause, but that he died so soon that he could do no more for it'.[2]

Oliver Cromwell led the Parliamentary forces, but the unpopular execution of King Charles I on 30 January 1649 brought England close to anarchy. Cromwell responded by consolidating his strength and ran a military dictatorship more or less successfully until his death in 1658. Once again, the country

The open forecastle bulkhead, with no panels, is characteristic of models from the reign of Charles II. The bell is housed under an impressive carved and gilded belfry canopy. The camber of the forecastle deck beams is evident, and young lions peer out at the break of the sheer rails. The jeer capstan is of the drumhead type and is fitted with five whelps and ten bars.

Despite being fully armed, only one pair of cannon are visible on the exposed decks. These guns have simple wooden carriages without trucks and are quite typical of armament on models of this period.

faced bloody and prolonged civil war unless it could be united, and this could only happen under the banner of an acknowledged and acceptable leader. A return to monarchy, albeit with a stronger and more independent Parliament, was the quickest way to end the violence and restore order, and General George Monk led the effort to accomplish this. Charles Stuart, son of the decapitated King Charles I, was summoned back from exile in Europe and with great fanfare was crowned King Charles II at Whitehall on 8 May 1660.

Sir William's grandson, Robert Fairfax, served in the Restoration navy most of his adult life, rising to the rank of admiral and becoming a Lord of the Admiralty in 1708. Whether the model was a gift, a bequest, or a purchase is not known, as it pre-dates both his tenure on the Admiralty Board and his naval service, which began in 1688. The exact circumstances of its acquisition are consequently a matter of speculation, and as will be seen, so is the identification of the ship it represents.

With the restoration of the monarchy, England was able to turn its attention back to world affairs and embark upon a period of political and economic expansion that would create an empire. This process began inauspiciously with the conclusion of a series of trade wars with Holland, begun during the Commonwealth period, that occupied most of the rest of the century and were fought principally at sea. Challenges from Spain and France

A CHARLES II 5TH RATE c1860

This photo, taken over sixty years ago, shows the model as it looked when exported from England. The model had been misguidedly 'rigged' at some time in the last century. The height of the case is not tall enough for proper rigging; instead, the height is appropriate to accommodate poles for launching flags, which would have originally been fitted.

followed, but in the end, Great Britain emerged as the greatest sea power in the world, and the Restoration navy was largely responsible. The origins of this success can be traced to innovations in naval architecture and an ambitious shipbuilding programme conducted by Parliament during the Commonwealth and Protectorate Periods. During the decade of the 1650s British dockyards produced seventeen 3rd-rate and twenty-eight 4th-rate naval ships that set new standards for warship design.

EVOLUTION THROUGH NATURAL SELECTION

The ascension of Charles II to the throne brought a period of further innovation in warship design, especially in the smaller rates. Because naval ships in the second half of the seventeenth century carried 10 to 12 guns per side on each deck, a ship of around 30 guns would have to carry a battery and a half. Fifth-rate 'frigates' accomplished this by bearing an incomplete complement of guns on the upper deck. Eliminating guns in the waist allowed the main gun deck to rise higher off the water, enabling the firing of guns on both broadsides, even in strong winds. Several such 5th rates were built during the reign of Charles II, and this model shares with them its gun port arrangement.

The original carved supporting cradles with their winged cherub motifs, a possible reference to immortality, are unique in our experience. As decorative elements they nicely counterbalance the gilded work on the superstructure and emphasise the baroque embellishments of the stern.

These small 5th rates were poorly suited for a line of battle, and the Royal Navy used them for scouting and pursuit as well as for convoy and anti-privateering work. Many decades later the frigate would become an indispensable addition to the sailing navy roster in squadrons and solo engagements, but in Restoration times these frigate progenitors were indispensable adjuncts to the large battle fleets. Our model is most likely descended from these early frigate prototypes, but is not actually one of them. The dimensions of our model on a 1/48 scale are several per cent too long and too narrow to correspond to any of these ships. More telling, however, are constructional features of the model that are too advanced to permit dating to the early 1670s. Most prominent are the upright stem, which made its appearance in 1678, and the subsidiary female companion to the lion figurehead, which also appeared after 1678. This is too late for a gunless-waisted ship to have been built for the British Navy.

What, then, does this model represent? The first vessel commanded by Robert Fairfax was the *Bonaventure*, a 4th-rate ship rebuilt in 1683 at Portsmouth dockyard. She had a complete upper tier of guns, however, and would not have exactly resembled this model. A more likely possibility is that the model represents a private man-of-war, and in fact, a building boom for small privately owned warships did occur around 1680. Venture capitalists of the time were acquiring men-of-war to attack richly laden Spanish and Dutch merchant ships. A dockyard model of one such vessel survives in the collection of the NMM, and represents the *Morduant*, built by Captain Castle at Deptford in 1683. She was built for a consortium of private investors, who eventually sold her to the Royal Navy, and her size and appearance are similar to our model. Future research may allow a more definitive conclusion, but at this time we can say that this rare and enigmatic model probably represents a privateer designed for speed, and built for dangerous extra-legal trading, such as with the Spanish colonies in America, an activity prohibited by Spain but flagrantly violated by English merchants.

References

Fox, Frank, *Great Ships the Battlefleet of King Charles II* (London: Conway Maritime Press, 1980), p. 208.

Gardiner, Robert, ed, *Line of Battle, the Sailing Warship 1650–1840* (London: Conway Maritime Press, 1992).

Right: This model is fully armed with original wooden cannon protruding menacingly from every gun port. As is typical for models from the period of Charles II, the colour scheme is largely black, red and varnished wood. The unusual winged cherub support cradles are noteworthy.

A CHARLES II 5TH RATE c1860

CHAPTER 3

The *Coronation*, 2nd rate of 1685

~ Acquisition ~

WHEN WE WERE YOUNG ENTHUSIASTS but not yet collectors, we never imagined that we would acquire an Admiralty Board ship model. This changed when we encountered a collection of important Admiralty Board models misidentified as nineteenth-century copies. The *Coronation* is the model with which we began our collection, but it was not the first model we acquired. While Henry was a college student in Boston, he first saw a Napoleonic prisoner-of-war ship model made of bone in Samuel Lowe's shop on Charles Street. It was far too expensive for him to buy, but it sparked his interest, and we began to study other examples and seek them out in museums, antique shops and auctions. We followed Christie's and Sotheby's furniture sales in London, which is where ship models would occasionally come up for sale, and we became known to the respective personnel in New York, who could let us know of any good models sold in the US, as there were no specialised marine sales in those days. Over the next several years we saw dozens of examples, but Henry's favourite was one that he found in the interior decorating department of Jordan Marsh, a department store in Boston. It was acquired when the store purchased the entire estate of Kenneth Roberts, Maine author and antique collector. It represented the English frigate *Pallas* and was unusual for its accurate proportions. It had already been for sale long enough that its price tag had yellowed with age, but inflation had not yet reduced its cost to be commensurate with Henry's budget. Not to be intimidated, he brazenly made a written offer for the model, which was politely rejected by the head of the department. No doubt to his later regret, this Jordan Marsh employee appended the words 'at this time' to his refusal. This phrase gave Henry great encouragement, and so he would repeat the same offer once a year for the next several years, hoping to find the 'right' time.

Henry had graduated and moved to California, where he was in medical school. In his second year, while back in Boston, he submitted his usual offer but this time to his delight, it was accepted. His surprise was tinged by embarrassment, however, when he remembered that even at this bargain price, he still could not afford to buy it. Henry called our father and explained

Left: This view captures the distinctive rigging of ships in the days of the spritsail topmast. The original rigging having deteriorated, the model was improperly rigged in the nineteenth century. It was re-rigged correctly by R C Anderson and L A Pritchard in 1920 while the model was on public view in Kensington Palace.

The poop and quarterdeck bulkheads are robustly carved and decorated. On the quarterdeck bulkhead, a drum, helmet and field gun are carved above the lion-draped canopy, marking the entry to a flight of stairs. At both quarterdeck and poop levels, there are 'twist' stairs with decorated gangways. The elaborate entry port is visible, with crouching lions draped over the arch, as on all the doorways on this model, and a pierced foliate rail encompasses the entry platform, which is a grating. Also visible are the lion-decorated gun port lids and lion mask scuppers.

his predicament, and much to Henry's relief Dad agreed to write the cheque (see Chapter 33).

Our father was very pleased with this purchase when he finally saw the model, and so began a collaboration that has continued for over thirty years. We imagined a fleet of miniature bone prisoner-of-war models taking shape, but this was not to be. As fate would have it, Arnold, who was in New York, heard from our contact at Sotheby's that a collection of four models was going to be sold at a house sale on Long Island on 30 May 1974. None were bone models, but the provenance was respectable, and Sotheby's thought we might

This port view of the quarterdeck shows both the stairs from the poop deck gangway and the admiral's staircase. This winding staircase leads to the bulkhead of the great cabin and is rarely fitted on models. Operation of both the adjacent quarterdeck gun and the main deck gun beneath this stair would have obviously been compromised.

be interested anyway. Photos were sent to Arnold, who forwarded them to Henry. There were four models, described as nineteenth-century productions representing earlier ships, and none were illustrated in the catalogue. From the black and white photographs

The quarter pieces take the form of cupidons riding on the backs of dolphins continuing the nautical motif displayed in the taffrail. The cipher of James II appears on the quarter gallery surrounded by lights made of mica and scored to represent panes of glass. The lower finishing takes the form of a winged cherub, and garter stars, emblematic of the order of the garter, adorn the topside frieze planking.

they sent us, we decided that Sotheby's was mistaken and that all four were period examples dating from the seventeenth and early eighteenth centuries. Besides the usual visual clues, our conclusion was supported by the provenance. These models were being sold from the estate of Junius Morgan Jr, grandson of J P Morgan and a contemporary of Henry Huddleston Rogers. Rogers, who had lived further east on Long Island, managed to collect over forty original Navy Board models, and it did not seem likely that Morgan, collecting at the same time, would buy four reproductions. We were so convinced by this reasoning that Arnold and our father attended the auction and bid for the model that we felt was the most important – a seventeenth-century rigged three-decker. The Morgan sale was held on the premises of Salutation, the Morgan estate in Glen Cove, Long Island, and lasted for four days. The ship models were sold on the second day, in the fourth session. This was an estate sale of a kind rarely seen in the US, and every seat in the vast outdoor tent was taken, and even the local television news team was in attendance. When the first model came up, and the bidding began to climb, the crowd became eerily quiet, or so it seemed to Arnold, who was doing the bidding. The bids quickly eclipsed the misguided estimate. The

A tall and regal rampant lion forms the figurehead with a naked cherub clinging to his mane. Of interest, the shipwright has made the foremost lower deck gun port wider than usual in order to allow the cannon to achieve a greater forward range of fire.

The forecastle bulkhead is decorated with carved and gilded caryatids, as was the custom, with particularly grand examples supporting the belfry canopy. The latter figures are depicted crouching in a rather awkward pose. The canopy is surmounted by two watchful lions beautifully contrived to conform to the arch of the belfry, and a metal bell with clapper has been fitted as if ready to strike. The belaying points along the rail take the form of carved heads, an early example of this practice. The jeer capstan is fitted with ten bars.

public enjoy the spectacle of a duel between determined bidders, especially when the price climbs well beyond what are assumed to be reasonable bounds established by the auction house estimates. By the time the hammer came down and ours was the winning bid, the bidding climax was met with loud sustained applause. More than thirty years have passed since that day, but the sense of excitement and satisfaction is as fresh in Arnold's mind today as it was during those five or ten minutes. It seemed at the time that everything in that sale was extremely expensive, but in hindsight most were bargains.

The day following the sale, Henry went to the Stanford University transportation library and found the model illustrated in *Contemporary Scale Models of Vessels of the 17th Century*, written by Henry Culver in 1926.[1] This confirmed that the model was a real example, and identified it as the *Coronation* of 1685. Of the four models, one of them, lot number 490, which was a mid-

eighteenth-century 74-gun ship, failed to sell at the auction. The following day we made an offer for it, which was accepted (see Chapter 14). Eventually we acquired all of the Morgan models sold that day, although it took eleven years.

~ Provenance ~

The original owner was John Vaughn, Earl of Carbery, who was Lord of the Admiralty during 1683–84. Upon his death in 1713, the model was acquired by Sir Richard Gough Kent and remained

THE *CORONATION*, 2ND RATE OF 1685

The high stern of restoration warships had given way, by this time, to the more compact, less vertical form visible here. The sheer at bow and stern is also reduced compared to earlier three-deckers, creating a slightly sleeker appearance.

Above: The cipher of King James II is proudly displayed in the centre of the taffrail, in a shield flanked by cherubs balanced on the heads of dolphins. Additional cherubs are riding on fanciful hippocamps to complete the lunate form. The royal coat of arms does not dominate the stern, as it did in earlier ships, but is here displayed in a robustly carved panel below the upper row of lights. There is an elegant recessed open gallery at the upper deck level, decorated with pierced foliate carved panels between the stern timbers, which are themselves adorned with a row of shy cupidons.

Right: The graceful lines of the underwater hull of this model suggest a ship that sat well in the water. The relative paucity of forward-firing guns underscores the inability of these ships to attack or defend against an enemy dead ahead.

at Gough House, Chelsea, until 1911, when it was lent to the London Museum, Kensington Palace, along with the model of the *Marlborough* (see Chapter 7). In February 1924, Mrs Anstruther Gough Calthorpe sold both models to the King Street antique dealer Rochelle Thomas for £1,400. It was her tacit understanding that they were 'not for the USA'. Nevertheless, they were both sold to New York dealer Max Williams who, in turn, brought them to his Madison Avenue shop and sold them to Junius S Morgan, Jr. We bought the model at an auction of his estate in 1974.

Description

CONDITION

This is model number 25 in Henry Culver's book, *Contemporary Scale Models of Vessels of the 17th Century*. Of those models that were rigged in the seventeenth century, only one, a model of a 4th rate still on display at Wilton House near Salisbury, England, has survived with the majority of its original rigging intact. Most seventeenth-century models have been re-rigged or have lost their rigging altogether. By the time the *Coronation* was lent to the London Museum in 1911, its original seventeenth-century rigging had been replaced with a nineteenth-century interpretation. The noted ship model scholar and author of *The Rigging of Ships in the Days of the Spritsail Topmast, 1600–1720*, R C Anderson, was concerned that the public would be misled by this erroneous display, and offered to correct it at his own expense.[2] This generous offer was accepted, and R C Anderson re-rigged the model in 1920.

The masts, tops and most of the yards are original, as are the guns and fittings including the spherical stern lanterns. The originality of these features was challenged in 1930 and authenticated by no less an authority than R C Anderson himself. Captain H Percy Ashley described the *Coronation* model in an article entitled 'A Noteworthy Shipmodel' that appeared in the September 1930 issue of the American *Shipmodeler* magazine. There he stated that, 'Most of the exquisite and artistic carving of the model is original and is very beautiful. It has all the charm of age and shadow that an expert could desire. It was renovated and re-rigged under the direction of Mr R. C. Anderson of London, who did quite a creditable job, although the guns and some of the metal fittings are not up to the original standard.'[3] This prompted R C Anderson to write, in a review of *The Shipmodeler* appearing in the British *Mariner's Mirror* in 1931, 'With regard to the model of the *Coronation* it is said that "the guns and some of the metal fittings are not up to the original standard." As a matter of fact, the guns are original and are exactly similar to those in several other models of the period.'[4]

CONSTRUCTION

Scale: 1/48 Hull length: 44½in

This model is one of only a handful of rigged seventeenth-century three-deckers that survive. The hull is framed in pear wood, which has acquired a golden brown colour. Unusual features include a row of scuppers below the upper wales that are modelled as lion mask gargoyles, and a lion's head painted on the inside of every gun port lid to further intimidate an enemy brave enough to approach at close range. The foremost gun port on the lower deck is considerably wider than the rest, which would give this bow-chaser an increased arc of fire. Ornate entry ports with crouched lions carved over the canopies are present on both port and starboard middle decks, which is unusual since most 2nd rates and many 1st rates merit only a single port-side entry. Winding staircases lead from the side gangways of both the poop and quarterdecks, but there is additionally an unusual twisted covered staircase leading from the quarterdeck to the upperdeck under the gangway on the port side only. Entry is through an elaborately decorated arched companionway.

Prior to the invention of the steering wheel in the early

This seventeenth-century cannonball, bearing the Royal Navy broad arrow mark, was recovered by Henry from the shipwreck lying on the sea floor.

eighteenth century, ships were steered by means of a whipstaff. This is the only known model of a seventeenth-century three-decker that has a whipstaff steering mechanism fitted, and this example is complete with a pivoting rowle through which the long whipstaff passes, the lower end of which is linked to the tiller via a metal crank. The helmsman was positioned on the middle deck, just abaft the mizzen mast in the windowless steerage, where he could have

A tall and regal rampant lion forms the figurehead, with a naked cherub clinging to his mane. Of interest, the shipwright has made the foremost lower deck gun port wider than usual in order to allow the cannon to achieve a greater forward range of fire.

no clue what course the ship was on. He was dependent upon instructions from the captain or master, two decks above.

The profusion of gilded carving on this model has been

compared to the model of the 1st-rate *Prince* of 1670 at the Science Museum in London. Notable elements include several large crowned monograms of James II along with other Royal insignia, a robustly carved trophy of arms including a wheeled field gun carved on the canopy leading to the main deck stairs, garter stars on the topside frieze planking and catheads and a magnificent lion figurehead.

Literature

The following references include photographs and descriptions of this model:

Anderson, R C, 'Books', *Mariner's Mirror*, 17 (1931), p. 420.

Anderson, R C, 'Comparative Naval Architecture, 1670–1720', *Mariner's Mirror*, 7 (1921), pp. 172–81.

Ashley, Captain H P, 'H.M.S. Coronation', *Shipmodeler* (September 1930), pp. 160–62.

Beach, Laura, 'Spreading Canvas', Antiques and the Arts Weekly, Newtown Ct, The Bee Publishing Co, 4 November 2016, p. 1.

Brown, C, 'Down to the Sea in Ship Models', *Forbes*, 144, 11 (13 November 1989), pp. 336–40.

'Coronation Ahoy!' *The Literary Digest* (24 May 1924), p. 31.

Culver, Henry, *Contemporary Scale Models of Vessels of the Seventeenth Century* (New York: Payson & Clarke Ltd, 1926), p. 25.

Davies, J D, *Pepys's Navy, Ships, Men & Warfare 1649–1689* (Barnsley: Seaforth Publishing, 2008), p. 43, 71.

Fox, Frank, *Great Ships the Battlefleet of King Charles II* (London: Conway Maritime Press, 1980), p. 164, plate 35.

Franklin, John, *Navy Board Ship Models 1650–1750* (London: Conway Maritime Press Ltd, 1989), cover illus., pp. 6, 23, 25, 36, 41, 42, 48, 57, 181, colour plates 2 and 3, 102–6.

Gardiner, Robert, ed, *The Line of Battle: the Sailing Warship 1650–1840* (London: Conway Maritime Press, 1992), p. 16, 166.

Hughes, Eleanor, ed, *Spreading Canvas*, New Haven, Yale University Press, 2016. pp. 142–3.

Kobak, Laurence B, 'British Admiralty Model Collection,' *Sea Heritage News*, 4, 12 (1983), p. 6.

Kriegstein, Arnold and Henry, 'The Kriegstein Ship Model Collection', *Nautical Research Journal*, 27, 2 (June 1981), pp. 81–93.

Kriegstein, Arnold and Henry, 'The Kriegstein Collection of British Navy Board Ship Models', *Nautical Research Journal*, 38, 4 (December 1993), p. 219, plates 2 and 3.

Lavery, Brian, *The Ship of the Line*, Vol. 1 (London: Conway Maritime Press Ltd, 1983), p. 31, 44, 224.

Laughton, L G C, *Old Ship Figureheads and Sterns*, (London: Halton & Truscott Smith Ltd, 1925), pp. 73, 132–33, 185.

Lounsbery, E, 'From the Smart Shops', *Arts & Decoration*, 26 (November 1926), pp. 14–18.

McCormick, W B, 'Model of H.M.S. Coronation', *International Studio*, May 1924, pp. 152–153.

Nance, R Morton, *Sailing-Ship Models* (London: Halton & Truscott Smith Ltd, 1924), p. 70, and plate 37.

Nance, R Morton, *Sailing-Ship Models* (New York and London: Halton, 1949), p. 56, plate 38.

Paxton, A, 'Ship Models Good and Bad', *Antiques*, 29 (April 1936), pp. 159–62.

Pardy, Kary, 'The Maritime World Through Miniatures and Beyond', Journal of Antiques & Collectibles, Vol. XX No. 8, Sturbridge, Ma., Weathervane Enterprises Inc, November 2019, p. 39

Pritchard, L A, 'Model of "Victory" in London Museum', *Mariner's Mirror*, 9 (1923), p. 157.

Roth, Leah and Ilene, 'Mirror Image', *Motor Boating & Sailing*, 148, 4 (October 1981), pp. 34–39.

Stow, C M, 'XVII Century Scale Models in Illustration', *Boston Evening Transcript* (4 April 1926).

W S, 'A Register of Models', *Mariner's Mirror*, 3, 2 (February 1913), pp. 57–58.

Watson, Peter, 'The Booty of Penlee Point', *The Sunday Times*, (20 November 1977), p. 13.

Winfield, Rif, *British Warships in the Age of Sail 1603–1714* (Barnsley: Seaforth Publishing, 2009), p. 35.

Exhibitions

London, Kensington Palace, London Museum, 1911–24.

Greenwich, National Maritime Museum, *Ship Models from the Great Age of Sail 1600–1850*, 18–20 April 1996.

New Haven, Connecticut, Yale Center for British Art, *Spreading Canvas*, 15 September–4 December 2016.

Historical Perspective

ENGLAND'S GREAT LOSS BY A STORM OF WIND

By the end of the Third Anglo-Dutch War the English battle fleet was outnumbered by both of its chief rivals, the French and the Dutch. Furthermore, half of its ships were over twenty years old. Largely due to the efforts of Samuel Pepys, Parliament was finally persuaded to remedy this deficiency by financing an expansion of the Navy. On 5 March 1677, £600,000 was authorised for the construction of thirty new ships. One was to be a 1st rate of 1,400 tons, nine were to be 2nd rates of 1,100 tons, and twenty were 3rd rates of 900 tons. The 1st and 2nd rates carried 90–100 guns on three decks and were all produced by royal dockyards. These thirty ships formed the backbone of the British Navy at the close of the seventeenth century, and they established the pattern of naval dominance that lasted over 200 years.

This model may have been built in 1677 as a design for the entire class of 2nd rates, as no other similar model exists and it does accurately represent the original appearance of these ships. However, the model bears the cipher of King James II in three places, and since the *Coronation* was the only 2nd rate built during his reign, this identification seems reasonable. The *Coronation*, built by Isaac Betts at Portsmouth dockyard, was the last of the thirty ships to be launched. She spent several seasons sitting on the stocks for want of funds to finish construction. Samuel Pepys recounts that construction took over five years, and that 'above one hundred pounds was demanded by her builder for repairing the decays of her very keel, as she lay upon the stocks'.[5] Furthermore, Pepys was able to collect mushrooms the size of his fist from inside the half-completed hull. She was finally launched in 1685, the year of the coronation of James II, and she was the largest ship launched during his reign.

LIFE AND DEATH OF THE CORONATION

The *Coronation* fought bravely at the Battle of Beachy Head on 30 June 1690. A combined Anglo-Dutch fleet engaged the largest French fleet that had ever been put to sea. The *Coronation* was the flagship of Vice Admiral Sir Ralph Delavall, who led the Blue Division, and she saw hot action in the engagement, although the allies ultimately lost to superior numbers.

Throughout the seventeenth century, the Great Ships of the British fleet withdrew to the safety of the Thames and the Medway for the winter. On 3 September 1691, while endeavouring to bear up for Plymouth, the English fleet was overtaken by a storm that dismasted the *Coronation* and sent her to the bottom along with her captain, Charles Skelton, and over 900 crew members. Only nineteen survived in the ship's boat. The *Harwich*, 70 guns, was also wrecked, and the *Royal Oak* (74) and the *Northumberland* (70) ran aground but escaped. This tragic loss of ships and men was long remembered, and a song about the disaster could be heard in forecastles 200 years later. A version collected in Nova Scotia in the twentieth century titled 'England's Great Loss by a Storm of Wind' includes the verses:

> *Twas on November the second day*
> *When first our Admiral bore away*
> *Intending for his native shore;*
> *The wind at west south-west did roar,*
> *Attended by a dismal sky,*
> *And the seas did run full mountains high.*
>
> *When we came to Northumberland Rock*
> *The* Lion, Lynx *and* Antelope,
> *The* Loyalty *and* Eagle *too,*
> *The* Elizabeth *made all to rue:*
> *She ran astern and the line broke,*
> *And sunk the* Hardwick *at a stroke.*
>
> *Now you shall hear the worst of all:*
> *The largest ships had the greatest fall.*
> *The great* Coronation *and all her men*
> *Were drowned except nineteen;*
> *The master's mate and eighteen more*
> *Got in their long boat safe on shore.*[6]

Surprisingly, the story of the *Coronation* doesn't end with her sinking. On 10 August 1977, retired British naval lieutenant Peter McBride and a team of amateur scuba divers got a strong magnetometer reading from the seabed ¾ mile off Penlee Point outside Plymouth harbour. They had been looking for the wreck

THE CORONATION, 2ND RATE OF 1685

of the *Coronation* for three frustrating years and had finally found it.[7] The wreck of the largest Royal Naval sailing ship ever found in British waters was lying in 70ft of water in easy sight of land. The site was identified by the recovery of a pewter plate made in 1689 and bearing the heraldic crest of the Skelton family. Charles Skelton was the unfortunate captain who went down with the ship. When Peter McBride's sub-aqua club learned that we had the original model of the *Coronation*, and that we were certified scuba divers, they invited Henry to dive on the wreck and even put his name on the official list of those authorised to salvage artefacts. Henry was finishing medical school in California but found time to travel to Plymouth with our father, and on one of the clearest days of the season, he visited the final resting place of the *Coronation*. Henry will never forget the thrill of descending through the murky waters of the English Channel until, at about

The *Coronation* was the only 2nd-rate ship launched during the reign of James II. She was the last of the thirty ships of the 1677 building programme to be completed. This port profile view shows how the hull and rigging are finely balanced both architecturally and aesthetically.

50ft, he could suddenly see the bottom, and it was littered with huge iron cannon barrels looking like giant pick-up sticks. At one-eighth the cost of brass, iron was the specified material for all the ordnance carried by the thirty ships of the 1677 Act, and while it decays quickly on exposure to air, the guns and anchors survived very well underwater. The timbers had, however, long since rotted away. An eel living in one of the cannon was about the only sign of life. There were also mounds of cannon balls, and Henry retrieved one that we still have as a tangible connection to the full-sized namesake of our model.

CHAPTER 4

The *Adventure*, 5th rate of 1691

Acquisition

THE BAROQUE CURVES, GILDED embellishments and rich wood tones of seventeenth-century models have always had a special appeal for us, with their magical ability to conjure up the days of Restoration England and bring us closer to notables of the period including Samuel Pepys and the van de Veldes. They are also rare as proverbial hen's teeth, which adds to their allure, and we are proud to have six in our collection. We first learned of this model when it came up for sale in Glen Cove, New York, in 1975. We did not buy the model at the sale, nor from the Manhattan dealer who did buy it. Years later, long after we had acquired the other three models sold from the Morgan collection, we succeeded in adding this last one to our collection through a trade for other naval art and artefacts.

Provenance

The early ownership of this model is not known, but it was in the United States by 1924, and was sold at auction on 27 March that year by the Anderson Galleries of New York, in its original walnut glazed case. It was erroneously identified as the yacht *Mary*, and by this time, the original lion figurehead had been replaced by a wax effigy of a saint. It was purchased by Max Williams, a marine antique dealer in New York City. He sold it to Junius Morgan Jr, who displayed it in his home in Glen Cove, Long Island. It was sold by Sotheby's at a house sale dispersing the contents of the estate on 30 May 1974, and was acquired by Landrigan & Stair, antique dealers in New York City. They, in turn, sold the model at auction in London, where it was purchased by an American naval historian. We obtained the model in a trade in 1986.

Left: The lines at the bow are fine, producing a slim profile. There is considerable sheer visible in the stern, quite typical for this period. The black upper frieze planking is made of ebony. The plinth and cradles are new.

Description

CONDITION

This model is No. 35 in Henry Culver's book, *Contemporary Scale Models of Vessels of the 17th Century*. When acquired, the original figurehead, decking and stern lanterns were missing. These were restored for us by Philip Wride, who had inherited tools and miscellaneous model parts from Robert Spence. Spence was an English collector and model builder who restored several Admiralty models in the mid-twentieth century. By great good fortune, among the bits and pieces Philip received was a cast of the original lion figurehead of the *St Albans*, a 4th rate of 1687 once in Spence's collection. This formed the basis of Philip's reconstruction for the *Adventure*'s figurehead. Lion figureheads were very standard at any given period, and this replacement must be a near-perfect match for the missing original.

Throughout most of the seventeenth and eighteenth centuries, a rampant lion was the standard figurehead on all but the largest British warships. While the overall design elements of this creature differed little from ship to ship; sometimes the beast appears fierce and noble as here, at other times more docile and friendly, much like a contented pet.

The trailboard behind the lion's foot takes the form of a stemmed Tudor rose. The iconography of the seventeenth-century cathead supporter is a mystery. Often, as in this example, a diminutive female torso is depicted with a male head and cloven feet.

CONSTRUCTION

Scale: 1/48 Hull length: 34in

The frame timbers are butted to the stem, keel, and aft deadwood on the starboard side, but the rising timbers are let down 1/16in into the deadwood on the port side. The floors also butt against a tall keelson, which rises above the level of the futtock heels. Flat floor riders are fitted athwartship, and paired footwales run fore and aft to strengthen the scarph of the futtocks with the floors. The channels are supported by timber spurs, and the short chainplates are single bolted to the chainwales.

There is a square tuck stern, which was common on Dutch ships, but unusual on a British naval vessel. This feature is, however, also seen on other small two-deckers of the William and Mary period. Additionally, quarterbadges are fitted rather than galleries, and both these unusual features also appear on two similar British models in the Royal Naval Museum in St Petersburg. Peter the Great visited England in 1697/8 in order to learn more about British shipbuilding practice, and a collection of models was given to the Czar to bring back to Russia. Among these are two vessels of similar date and size to this one, which also bear quarterbadges and have square tuck sterns. It seems reasonable to

The graceful lines of this snug two-decker are evident here. The belfry with its suspended bell is visible in the forecastle bulkhead.

speculate that these three models, and the ships they represent, may be somehow related.

The arms of William III appear in a tiny but distinct carving in the centre of the upper counter, flanked by reclining figures of female trumpeters. There is a carved bust of a lion in the middle of the taffrail with his right paw on a shield, surrounded by a wreath. Dolphins are entwined on either side. The quarter pieces consist of Roman warriors, and the port wreaths are carved with ribbons and Tudor roses. The lower finishing of the quarter gallery is a robustly carved cornucopia.

The model retains its original seventeenth-century display case, consisting of an oak carcass with figured walnut veneer, glazing and brass fittings, with a hinged and locked front panel. It closely resembles one of similar date from the Sergison collection now at the Naval Academy Museum in Annapolis. The cradles, plinth and stand are new.

Literature

The following references include photographs and descriptions of this model:

Culver, Henry, *Contemporary Scale Models of Vessels of the Seventeenth Century* (New York: Payson & Clarke Ltd, 1926), p. 35.

Davies, J D, *Pepys's Navy, Ships, Men & Warfare 1649–1689* (Barnsley: Seaforth Publishing, 2008), pp. 54, 71, 135.

Kriegstein, Arnold and Henry, 'The Kriegstein Collection of British Navy Board Ship Models', *Nautical Research Journal*, 38, 4 (December 1993), pp. 219, 220, and plate 4 and 5.

Philbin, Tobias, & Endsor, Richard, *Warships for the King* (Florence, OR: SeaWatch Books LLC, 2012), p. 124.

Winfield, Rif, *British Warships in the Age of Sail 1603–1714* (Barnsley: Seaforth Publishing, 2009), p. 167.

This model is preserved in a beautiful walnut-veneered display case.

Historical Perspective

The dimensions of this model match those of the *Adventure*, a large 5th-rate ship of 44 guns rebuilt at Chatham dockyard in 1691. Dimensions and scales of Admiralty Board models cannot always be trusted, however, and if this one was off by less than 2 per cent, it could fit the *Dragon* rebuilt at Deptford in 1690. Support for this contention comes from an interpretation of the carving at the centre of the taffrail, where proprietary designs are often found. In this case, there is the bust of an animal with fangs, a canine muzzle, porcine nose and leontine mane. In seventeenth-century iconography, this animal resembles a dragon. But until further evidence comes to light, we have decided that this model more likely represents the *Adventure*. Whichever ship she is, both vessels saw action against the French in the war that followed the 'Glorious Revolution' of 1689.

THE 'GLORIOUS REVOLUTION' INCITES A FAILED COUNTER-REVOLUTION

The Third and final Anglo-Dutch War had been concluded at Westminster on 9 February 1674. Commercial ties between the former enemies were established almost immediately, and the marriage of Princess Mary of York to William, Prince of Orange, elevated the relationship of the two powers to that of an alliance. Their natural enemy was, of course, Catholic France under the ambitious Louis XIV. Charles II, a Protestant, was succeeded in 1685 by his Catholic brother, James II, who enjoyed little popular support. Seizing an opportunity to put a Protestant, namely himself, back on the throne, William left Holland to invade England on 20 October, landing at Torbay on 5 November 1688. The Prince of Orange was welcomed in his adopted land, and a Protestant monarchy was restored, under William and his wife Mary.

The exiled King James II received support from his Catholic ally Louis XIV, and in an effort to return him to the throne, a French fleet landed James II at Kingsale in Ireland on 12 March 1689. In the doomed counter-revolution that followed, the French navy fought against England and her ally, Holland, in a series of actions including the battles of Bantry Bay in 1689 and Beachy Head in 1690. No fleet engagements occurred during the 1691 season, but by 1692 the French were preparing an invasion force. The Allies assembled a powerful fleet to thwart this effort, and on 19 May they engaged a numerically inferior French fleet off Cape Barfleur. The *Adventure*, Captain Thomas Dilkes, fought in the Blue

The deadwood is unusual as being fashioned from a single piece of boxwood. There are no accommodations shown for foredeck or quarterdeck guns.

THE *ADVENTURE*, 5TH RATE OF 1691

An unusual feature is the square tuck at the stern, more commonly found on Dutch ships of this period. This model has a quarterbadge typical for smaller two-deckers, but in the model this is carved of one piece of wood including the partially open port and its hinges. A robust cornucopia extends below the quarterbadge.

Squadron. The outcome was a foregone conclusion. When twelve large French men-of-war took refuge in the Bay of La Hogue, Vice Admiral Rooke sent fireships in and burnt them all in view of ex-King James, whose hopes for restoration to the throne were extinguished with the flames.

In October that year, the *Adventure*, cruising with the *Rupert* off the Irish coast, fought and captured two privateers along with their two prizes and two merchantmen. Later in December, the *Adventure*, still under Captain Dilkes, captured two 16-gun privateers, and in May 1694, she assisted in the capture of the *Diligente* (36).

In 1695 the *Adventure* was in the Mediterranean and captained by Charles Cornwall. On 7 January, she was in a small

The taffrail is unusual because it is asymmetric with the beast holding a shield to the port side. The quarter pieces are sword-brandishing classical warriors and would have been intimidating by their size on the real vessel. The royal arms are here reduced to a relatively modest escutcheon on the upper counter, unaccompanied by lion and unicorn.

squadron commanded by James Killigrew in the *Plymouth*, 60 guns. When Killigrew encountered the French ships *Content* (60) and the *Trident* (50), he attempted a time-honoured ruse and hoisted French colours. The French ships, for their part, hoisted English colours. Despite these pretensions, a hot engagement ensued during which Killigrew was killed, but the English squadron prevailed, capturing the two French vessels. The war was concluded by the Treaty of Rijswijk on 11 September 1697, when France accepted King William and denounced ex-King James.

References

Clowes, Wm Laird, *The Royal Navy a History From the Earliest Times to the Present*, Vol. 3. London: Sampson Low, Marston and Company Ltd, 1898.

CHAPTER 5
A William III 4th rate c1695

~ Acquisition ~

MANY OF THE MODELS in our collection found their way to America in the first decades of the twentieth century, most to enter the collections of American captains of industry of the time. However, during the Great Depression of the 1930s the fad for ship models faded quickly and their value declined. As models passed from generation to generation their identity and importance was often forgotten. This was a great advantage to us when we began collecting. We first became aware of this model as a result of its inclusion in Henry Culver's book, *Contemporary Scale Models of Vessels of the 17th Century*. In that catalogue, the model was said to belong to Colonel H H Rogers. A thorough search of the Naval Academy museum at Annapolis, the final repository of Col Rogers' collection, confirmed that this was an erroneous attribution and the model was never a part of that collection. It became, for us, a 'missing' model, and we kept our eyes peeled in hopes of finding it someday. It wasn't long before we did, in fact, 'discover' the model on display in the collections of the Mystic Seaport Museum in Mystic, Connecticut. The upper works of the model had been covered in thick black and gold paint, but the surface underneath was complete and well-preserved. It bore a museum label misidentifying it as a Queen Anne model and describing it as a gift from Clarkson A Collins Jr of Rhode Island. We were pleased to have solved the mystery of the whereabouts of this important model, but at the same time were disappointed that it was no longer in private hands. Nevertheless, we regarded this information in a positive light as offering the prospect of new discoveries. We reasoned that the Collins' family might have retained other items, perhaps even models, which were not given to the Mystic Museum, particularly because the mission of that museum is to preserve the maritime culture of America and New England, and not the British Navy.

We quickly found a telephone listing for the Collins family and spoke with Clarkson Collins' son. Much to our delight, he did, in fact, still have a number of models that his father had bought or built. Mr Collins was most gracious and kind, and invited us to visit and see what he had. Our excitement at the prospect of finding a trove of Navy Board ship models was dampened by the reality of finding that none of the models in the house were dockyard examples. There were about ten models, and although they formed

With the exception of the lower shrouds, the rigging dates from the twentieth century and was added while the model belonged to Clarkson Collins Jr. Mr Collins once offered to sell the model to Colonel Henry Huddleston Rogers, but the sale was never consummated.

The arrangement of the forecastle and beakhead are shown in this illustration. Note the gilded main headrails terminating in carved knightheads. This model pre-dates the introduction of roundhouses on two-deckers.

a diverse collection and were of some interest, we did not offer to purchase any of them. Mr Collins was also visibly disappointed, and asked us what kind of model we were looking for. We replied that we were hoping to find a model like the seventeenth-century one his father had given to the Mystic Seaport Museum. 'What about *that* one?' he asked. We did not understand his meaning and explained that the model had been given to the museum and was not available. He disagreed. He insisted that the model had been *loaned* to Mystic, and even went so far as to complain that in all the intervening years, we were the first ones to express any interest in it! Although we left it for him to sort out the true state of affairs with regards to this model, we couldn't resist a call to the museum registrar upon our return from Rhode Island. To our enormous surprise, the label on the model was in error, and this incredible survival from the late seventeenth century was, indeed, still in private hands! Because the title to the model belonged to an uncle who was in Portugal, it took nearly a year to conclude the purchase.

Provenance

The early history of this model is not known, but it was acquired by Clarkson A Collins Jr of Providence Rhode Island in the early part of the twentieth century. On 10 August 1922 he wrote a letter offering this model, along with two others, to Colonel Henry H Rogers for $15,000. Colonel Rogers did not buy this model. It was lent to the Mystic Seaport Museum following the death of Clarkson Collins and was purchased by us in 1976.

Description

CONDITION

In the letter to Colonel H H Rogers in 1922, Mr Collins described the model as in 'perfect condition', and although the running rigging and greater part of the standing rigging had rotted away, it still retained 'all of the original masts, round-tops, and yards'. When this model was acquired in 1977, all the carvings were covered with thick gold paint and many other surfaces, including the mica windows, were painted black. All the overpaint was carefully removed during conservation and cleaning in 1988 to reveal the original gilded and varnished surfaces underneath. The model was

This is a view of the model undergoing a cleaning in 1988. Modern black and gold paint was removed from the decorated surfaces, exposing the original gold leaf and ebonised finishes.

originally fitted with a tilt frame for an awning over the quarterdeck, and while the stanchions that held it in place remained, the original frame disappeared sometime between 1936 and 1952. In September 1988 an exact copy of the original frame was made by Philip Wride based on photographs of the original as it appeared on the model in 1926. Over the course of three centuries the mizzenmast developed a warp, causing it to bend forward. Rob Napier was able to correct this distortion by adjusting the tension of the rigging on the mast.

CONSTRUCTION

Scale: 1/48 Hull length: 35in

This is model number 36 in Henry Culver's book *Contemporary Scale Models of Vessels of the 17th Century*. It is an unidentified 4th-rate ship and bears the cipher 'RWR' for William III. The absence of an 'M' for Queen Mary and the presence of a gilded bust of William in a cartouche on the starboard side of the stern without a

The masts, spars and shrouds on this model are original, as are all the carvings including the cradles.

The cradles of seventeenth-century models are often carved as dolphins, but these are particularly fine and elegant examples. Nearly every available surface has been decorated. As a result, it is difficult to see that hinged gun port lids have been cut into the frieze of the gallery.

corresponding carving of Mary to port, suggests the model was built after Mary's death in 1694. The hull is very carefully constructed, and an especially handsome feature is that the lower edge of the scarph between the futtocks and floor timbers runs ¼in above the top of the deadwood and parallels the curve both fore and aft.

    The carving is especially delicate, and besides the gun port wreaths, knightheads, figurehead and terms at the head and breaks of each deck, the stern is beautifully decorated with crowned astrolabes, lions' masks and winged cherubs. The upper counter of

the stern contains William III's coat of arms in a central panel flanked by stern chaser port lids that are nearly concealed by finely executed foliate carvings. The monogram of William III appears beneath the stern lights, and in the centre of the taffrail there is an intriguing carving of a bust of a horse flanked by amorini and dolphins. The animal appears to be a sea horse because there are webbed hooves on the projecting forefeet, and the presence of the sea horse may relate to the name of the ship. If so, we have been

unable to account for it. The stern is also embellished by gilded painting, including thistles below the lights on either quarter gallery and trophies of arms on the lower counter. The mica panels on the stern lights feature elaborate engraving to represent the pattern of leaded joints in the full-sized glass panes.

Literature

The following references include photographs and descriptions of this model:

Culver, Henry, *Contemporary Scale Models of Vessels of the Seventeenth Century* (New York: Payson & Clarke Ltd, 1926), p. 36.

Culver, Henry, 'Private Collections of Ship Models', *Antiques* (September 1923) pp. 125–31, and see frontispiece illus.

Franklin, John, *Navy Board Ship Models 1650–1750* (London: Conway Maritime Press Ltd, 1989), pp. 33, 42, 60, and colour plate 7.

Kriegstein, Arnold and Henry, 'The Kriegstein Ship Model Collection', *Nautical Research Journal*, 27, 2 (June 1981), pp. 81–93.

Kriegstein, Arnold and Henry, 'The Kriegstein Collection of British Navy Board Ship Models', *Nautical Research Journal*, 38, 4 (December 1993), pp. 220–221, plate 6, and cover illus.

Nautical Research Journal, 13 (1965): frontispiece, illus.

Philbin, Tobias, & Endsor, Richard, *Warships for the King* (Florence, OR: SeaWatch Books LLC, 2012), p. 160.

Stow, C M, 'XVII Century Scale Models in Illustration', *Boston Evening Transcript* (4 April 1926).

Exhibitions

Mystic, Connecticut, Mystic Seaport Museum, 1942–76.

Left: Photographs of this model taken prior to 1922 document the original tilt frame over the quarterdeck, which unfortunately had been lost. However, the wire stanchions holding it in place all survive and we had a replacement made for the missing frame.

Above: The graceful upward taper of the stem can best be seen in this dead-ahead view. The delicate and lofty aspect of the rigging provides an aesthetic counterpoint to the hull and gives a visually balanced sculptural impact offset by the gilded cradles.

Historical Perspective

IDENTIFYING SHIPS AT SEA

We have often wondered how it was possible to identify men-of-war at sea in the seventeenth and eighteenth century. Logbooks of the period attest to the ability of seamen to recognise a specific ship even when sighted at a distance. The custom of painting ship's names on their sterns began in 1771 for Royal Naval vessels, and names were not displayed on the bows of ships until the mid-nineteenth century. How was the feat of recognition achieved in the days of the Restoration navy of the seventeenth century? Flags simplified the task. Flags at this period did not signal a ship's name, but in a custom that began in the previous century, ensigns, pennants, and flags were flown to distinguish the relative status of individual ships within fleets or squadrons. Fleets consisted of three squadrons, the Red, the White, and the Blue, distinguished by the colours of the flags they flew. The flagship usually flew the Union flag at the main, unless the Duke of York, who was Lord High Admiral was on board, in which case the Royal Standard was flown. The admirals of each squadron flew a plain flag in the appropriate red, white or blue colour at the mainmast, while vice admirals flew appropriately coloured plain flags at the foremast, and rear admirals flew a flag of the appropriate colour at the mizzen. Ships of each squadron flew appropriately coloured ensigns at the stern and Union Jack at the bowsprit. In small fleets in which the squadrons were not subdivided, the admiral flew a Union flag at the main, the vice admiral a Union at the fore, and the rear admiral a Union at the mizzen. All the ships in such small fleets wore the red ensign. The flags and their arrangement were adjusted when commanders or fleet assignments changed, and therefore were not reflective of the identity of an individual ship. Nonetheless, it was possible for a seaman to identify a specific vessel glimpsed at sea if that vessel enjoyed a unique position in a specific squadron at a particular time. For example, English officers knew in May 1672 that the vice admiral of the Blue Squadron flew his flag in the *Royal Sovereign*, that the rear admiral of the Red Squadron sailed in the *Royal Charles*, and the commander of the Blue squadron sailed in the *Royal James*. Sailors and officers on board other ships in the vicinity could identify these ships by the flags of command they flew.

Perhaps the most famous instance of this occurred during the Battle of the Texel, on 11 August 1673, the last naval battle of the Third Anglo-Dutch War. On that memorable day, Sir Edward Spragge commanding the *Prince* opposed Cornelis Tromp in the *Golden Lion* and in the fierce struggle that ensued, both flagships were so disabled that their commanders were forced to abandon their ships and shift their flags. Spragge transferred to the *St George*. In the terrible fight that recommenced, the *St George* was also disabled, and Sir Edward planned to shift his flag once again, this time to the *Royal Charles*. While transferring command, Spragge apparently carried his flag with him in his barge, a fatal mistake that drew the unwanted attention of the Dutch. A cannon shot swamped the barge. The sailors valiantly rowed back toward the *St George*, but as they groped for the ropes thrown to them by their shipmates, the barge slipped beneath the waves and Sir Edward, who could not swim, was drowned.[1]

Flags could provide critical information to help identify specific ships under certain circumstances, but one wonders if ships were also identified at sea by their most individual features, namely their gun port arrangements and decorative carvings. Aside from flagships of the 1st or 2nd rates that often had unique figureheads, most warships bore lion figureheads and could not have been distinguished from each other by sighting their bows. The most distinct features were the carvings, terms and mouldings of the sterns and quarter galleries, and these may have been familiar enough to seamen to allow identification even if glimpsed from afar. We are aware of no documentary evidence supporting or refuting this proposal, but it seems reasonable to us. Moreover, the notion that a sailor at sea might have glimpsed a distant ship through his spyglass and declared it to be a specific ship by recognising the carvings on its stern is appealing since these are among the features we most admire on the miniature versions.

Right: When we acquired this model the gilded and ebonised surfaces were covered in paint. Careful cleaning and conservation revealed the original patinated surfaces.

A WILLIAM III 4TH RATE c1695

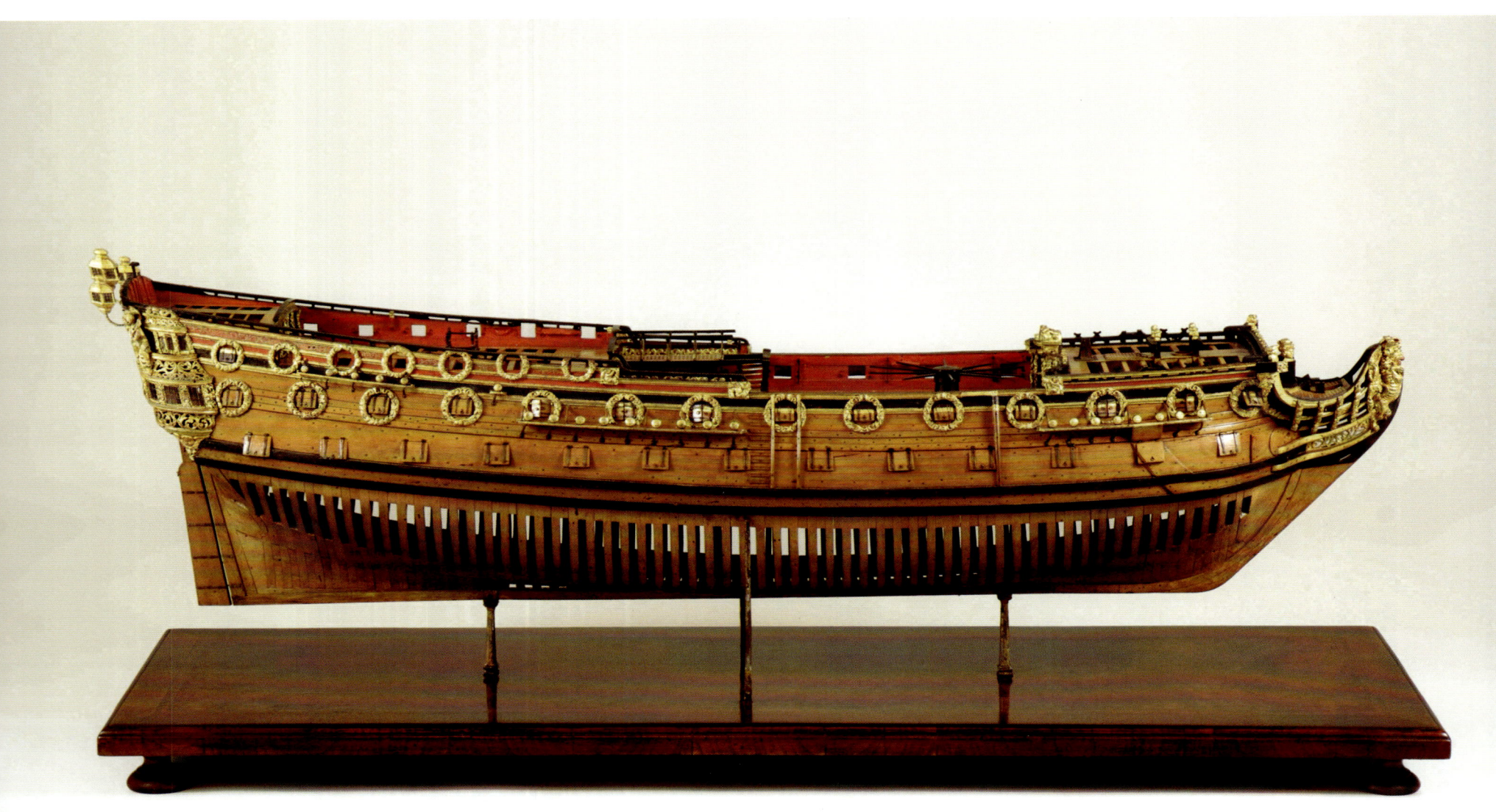

This model, built c1702, is adorned in seventeenth-century style, bearing carved and gilded gun port wreaths, head timbers, catheads, bulkheads, hancing pieces and elaborately decorated quarter galleries and stern, though with conspicuously fewer putti than on earlier ships. This was one of the last ships to be decorated in this seventeenth-century style, as the Admiralty order of 1703 placed severe restrictions on expensive decoration and ended the era of exuberant carving and gilding.

CHAPTER 6
The *Northumberland*, 3rd rate of 1702

~ Acquisition ~

FOR A FEW BRIEF years at the beginning of the eighteenth century, the elegant craftsmanship of the Queen Anne era was wedded with the baroque carving and gilding of the seventeenth century. Asking us which model is our favourite is much like asking a parent with a large family which child he likes best, but if pressed, this model would surely be on the shortlist for both of us. This beautiful model of a 70-gun ship, c1702, appeared at a Sotheby's auction in London in 1975. At that time, Sotheby's included models in their furniture sales, and this one appeared sandwiched between a bureau bookcase and a fine Sheraton table. It was the property of Sir John Molesworth-St Aubyn, Bt, CBE. We later learned from him that he was a descendant of Thomas Herbert, Lord Pembroke, who was Lord High Admiral from 1701–02. Presumably he received the model during his tenure on the Admiralty Board. Arnold flew to London to attend the sale. Our enthusiasm, which was considerable, was nonetheless tempered by the fact that the model appeared to have active woodworm. There were the usual numerous woodworm holes in some of the timbers, but there were also small piles of wood powder (frass) visible in several places. This was an early stage in our collecting activities, and we had not encountered woodworm problems before. Because we did not know how serious the problem might be, nor how to deal with it, we were cautious with our bidding. As it happened, we were the underbidders. The purchaser was the British Railway Pension Fund, although we only learned this years later. Sotheby's had been advising the fund on fine art investments and were selecting items of impeccable historical or artistic merit in order to help them develop an investment portfolio. The obvious conflict of interest did not seem to deter the fund managers. This model had attracted their attention, and they were determined to have it. The day after the sale, Arnold met with the curator of ships at the Science Museum, Dr Basil Bathe, about another matter, and Dr Bathe said that he had attended the sale and asked whether we had bought the model. Arnold replied that we had not, and that we thought the

The belfry canopy is uniquely decorated with the carved figures of two slumbering cherubs with an hourglass between them. The galley funnel is fitted directly forward of the belfry. The supporters of the belfry are decorated with caryatids of typical seventeenth-century form, while the rest of the vertical timbers of the bulkhead are relatively slender timbers decorated with foliate carvings. The timberheads of the beakhead bulkhead terminate as gilded knight's heads, smaller versions of the heads carved on the foremast bitts.

price was quite high. He agreed that it was indeed a lot of money, but after a contemplative pause he added, 'but one really does have something there'. Being the underbidder on a coveted lot is never a happy position, but under these circumstances, it was especially grating.

One of the things we have learned over the years is that patience can have its rewards, and seven years later the pension

The beautiful symmetry of the hull can be appreciated in this view. A seat of ease is visible tucked in the space between the main rail and the beakhead bulkhead. An unusual feature is the grating that forms the backrest for the seat. The top timbers of the beakhead frame are unusually prominent just above the seat, and one can imagine that many poor sailors struck their heads against this timber while answering nature's call.

The quarter galleries are compact and wrap around neatly to blend with the mouldings and friezes of the stern. The upper finishing of the quarter gallery takes the form of a serpentine dolphin, while the lower finishing is a winged cherub's head. French doors lead off the upper tier to a small balcony, and the breast rail with its pierced foliage decoration that might otherwise be taken for a canopy is actually the balcony rail.

fund decided to liquidate some of its investments and the model was sold again. It was in unchanged condition, having remained at Sotheby's in the office of the director all those years, and we were surprised to observe that the woodworm damage had not progressed. This time around, Henry and our father attended the sale, and we were very determined bidders. We were successful and well aware how lucky we were to have had an opportunity to replay the auction with a more satisfying result the second time around. But we were soon to learn that purchasing the model and bringing it home were two different matters, and the acquisition process had only just begun.

THE COMMITTEE ON THE EXPORT OF WORKS OF ART CONFRONTS A SHIP MODEL

Prior to 1983 our Navy Board models were all purchased in the US and, therefore, never had to deal with export issues. However, this model was the first one we purchased at auction in London. Certain works of art require an export licence to be legally exported from the UK. Except for manuscripts or items unearthed from British soil, which all require export licences, most other items such as paintings, sculptures or other works of art require such licences only

A grand sideways-facing bell staircase leads from a companion on the quarterdeck to the upper deck. Finely turned pillars supporting the deck beams are visible.

if they exceed a certain monetary value. This model qualified. Consequently, we duly applied for an export licence. Our application, consisting of a brief description with a photograph, wound up on the desk of Dr A P McGowan, the Head of the Department of Ships at the National Maritime Museum in Greenwich, who also happened to serve as the Expert Adviser on maritime objects for the Reviewing Committee on the Export of Works of Art. He opposed export because he deemed the model to be an object of National Importance, or more specifically, he claimed the model met two of the three Waverley criteria. The Waverley criteria, named for the chairman of a committee appointed in 1950 to advise on export policy, define the attributes of a work of art that qualify it as an object worthy of acquisition for the Nation. Fulfilling any one of three criteria will qualify an item for retention on the grounds of national importance, and the three criteria are: 'Is it so closely connected with our history and national life that its departure would be a misfortune? Is it of outstanding aesthetic importance? Is it of outstanding significance for the study of some particular branch of art, learning, or history?'[1]

Dr McGowan claimed that the model met the first and third criteria, and meeting either one was sufficient for refusal of an export licence. The model thus became the subject of a hearing before the Reviewing Committee.

The Committee consists of eight members appointed by the Secretary of State for Culture, Media, and Sport who have expertise in one or more fields (paintings, sculpture, furniture, manuscripts, etc.). In this case the committee was chaired by Lord Plymouth (who was hard of hearing) and included the keeper of the Queen's pictures, the curator of the Tate Gallery, the curator of the National Gallery, and a number of distinguished curators of other art museums from around the British Isles, including two nautical experts, a former Director of the NMM and a curator of the Merseyside Maritime Museum, who were appointees specifically added to the committee on this occasion to replace Dr McGowan.

When an export licence is referred to the Reviewing Committee, the applicant is invited to submit a written statement giving reasons why the object does not meet the Waverley criteria, while the Expert Adviser is asked to provide a statement supporting his or her view that it indeed does satisfy one or more of the criteria. The parties are then shown each other's statements, and they can prepare rebuttals. A meeting is convened within weeks, where the Committee, including up to three additional advisers chosen for their expertise relating to the object in question, can listen to verbal arguments, ask questions, and render a judgment. The statement prepared for the Reviewing Committee by the NMM was cogent and compelling. It made the case that while the Museum's holdings of contemporary dockyard models was second to none in size and importance, they lacked an example comparable to ours. It went on to conclude that this model must be considered of 'national importance'.[2] In order to keep the model, we had to convince the Committee otherwise. Our only chance to accomplish this was to present even more convincing counter-arguments in our rebuttal. We spent several weeks labouring over this, and then set out to London to make our oral delivery before the Committee.

We had no hope of success. We were certain this would be a hanging jury, but we were excited to be participants and were resolved to have a good time. To our surprise, the hearing was held at Whitehall, the old Royal Palace, site of the beheading of King Charles I, and directly across the street from the Admiralty offices. It is generally thought that Admiralty models were brought to the meetings held at the Admiralty in order to be examined by the commissioners prior to deciding on the year's building programme. Nearly a century ago, J Seymour Lucas RA painted an imaginary scene, 'The New Design', based on this concept. The painting is now in the Victoria and Albert Museum. It depicts a distinguished group of Navy officials gathered around a table to inspect a ship model while listening attentively to the presentation of a naval architect. This painting was subsequently turned into a diorama displayed in the Hall of Water Transport at the Science Museum, London, but further imaginative adjustments have been made since King Charles II and Samuel Pepys are now featured among the Navy officials. However, it seems unlikely that examination of ship models played a direct role in winning commissions for the dockyards as they are never mentioned in any accounts of Admiralty or Navy Board meetings. Moreover, we know that at least some models were built after the ships they represented were already on the stocks. Nonetheless, an invitation to bring our model to a hearing at Whitehall, to be scrutinised by an august body of art experts and to argue our case for export with the authorities from the NMM, proved irresistible. We also had no alternative if we wanted to bring the model home.

We argued that there were already over 150 models in the national collection, including several close in size and date to ours, and this particular one was not outstanding enough to merit retention. The museum argument was basically that every model was important, and they said something about the 'jigsaw puzzle' of shipbuilding where every piece was significant. We did not agree with this argument, particularly since the museum had not blocked the export of several other wonderful models that had left the UK in the recent past. Things, however, were not going well for us, and Dr McGowan took every opportunity to claim that we were misleading the committee by understating the importance of our model. Arnold saw a ray of hope when the museum conceded that the model might never be identified with a named ship since it may have been made as a proposal for a ship that was never built. In fact, the measurements did not exactly match any known ship of the period. But the critical moment came near the end of the proceedings, when the keeper of the Queen's pictures asked Dr McGowan how we were able to state that this model, with no date visible, was built in 1702, and also asked him to show or tell the committee what specifically was so important about this particular model that would merit retention. This drew a response about how any student of naval architecture could date a ship based on the presence or absence of features that changed gradually over time, and something about how this model showed what a 70-gun ship might have looked like around 1700.

We couldn't believe our luck! We looked at each other quite surprised at how inadequate this answer was. Arnold took the bold step of asking whether he could be permitted to give a response to these same questions. Lord Plymouth, seated exactly opposite us at the head of the immense table, held the hearing trumpet up to his ear and said: 'Eh?' Arnold then stood up and in a raised voice repeated his request to answer these two questions. The request was granted. Arnold was struck at how much this hearing seemed like theatre, and feeling like an actor in a play, he proceeded in his most stentorian voice. In a brief speech he pointed out that identified models of named ships, whose dates of construction were known, were the source of the timetable of design changes. Thus models of specific named ships had truly outstanding value. Models of unidentified, possibly never-built ships, could never reach the same level of importance, regardless of how marvellous or beautiful they may be. It must be remembered that at the time of the hearing this model had not been identified, and it is only now, forty-five years later, that we are proposing she is the *Northumberland*. This was the last statement of the hearing, as no further questions were asked, and the plaintiffs on both sides were asked to leave while the committee deliberated.

There were some tense moments in the waiting room. The Maritime Museum members were quite confident that the committee would vote for retention of the model, thus giving them three months to come up with at least half the purchase price and up to six months to raise the rest. Dr McGowan informed us that they had already placed a suitable display case for the model in the rotunda of the museum along with a plaque describing the urgency of preventing this important model from leaving the country and asking the public to contribute toward preserving the model for the nation. Of course, he confessed, a call to one or two key benefactors of the museum would probably suffice to raise the necessary funds, and Dr McGowan was confident that there would be no problem at all in that regard. We also expected this outcome, as we knew that this particular model was an unusually fine example with several interesting design features and could well qualify as a national treasure, but we were acutely aware that the museum had totally missed what these unique features were and had failed to make a compelling case for why this model was truly outstanding. Instead, Dr McGowan's approach was to go on the offensive and make

An innovative feature is the provision for removable gangways. Three pairs of L-shaped metal brackets are mounted along the bulwarks at the waist and fitted so that they can swivel. These brackets support a gangway's plank (not present), a feature that would permit rapid transit from the foredeck to the quarterdeck without the trouble of climbing two flights of stairs. It is surprising that such a practical innovation was not widely adopted until the 1740s.

several remarks attacking our list of comparable models already in Britain, which he claimed to be inflated and misleading. We were dismayed at the prospect of losing this argument along with our model, when in our view we had made a stronger case. In about fifteen minutes we were asked to re-enter the painted hall to hear the verdict.

Of course, the reader will already know the outcome, as this model is, after all, in a book about the Kriegstein collection. But we still recall our incredulity when Lord Plymouth announced that an export licence would be granted. The Committee then promptly adjourned for lunch. Several members approached us with congratulations, and at least one remarked about how beautiful he thought the model was. Dr McGowan seemed taken aback by the decision and left the room without a further word to us. As we later learned, he and one of the consultants for marine objects appealed the verdict on the grounds that the discussion had been abbreviated because committee members were eager for lunch. This appeal, however, was eventually denied. As we were preparing to leave, one of the museum representatives approached us to ask whether we would lend the model to the NMM for study as we had offered in our opening statement. We responded that we were ready to bring it there immediately, and while he kindly offered to arrange for a van, we simply took a taxi and delivered the model to the museum that afternoon. We found that the display case had already been set up in the rotunda to accept the model, as we had been informed, but the collection bowl and label needed to be removed. The model remained at the NMM for a total of six months and was photographed, measured and examined carefully.

Provenance

The original owner of this model was Thomas Herbert, Earl of Pembroke, Lord High Admiral in 1701–02, and again in 1707. It descended in his family until 1975, when Sir John Molesworth-St Aubyn Bt, CBE sold it at Sotheby's in London. The model was bought by the United Railway Pension Fund, who kept it for eight years until selling it again at Sotheby's in 1983, when we acquired it. The model was exhibited at the NMM, Greenwich, England, in 1983–84.

Description

CONDITION

Wormholes are common on seventeenth-century models, but shortly after this one was acquired, tiny piles of sawdust appeared beneath holes on the deck planking and hull timbers, indicating that live passengers were aboard. We had the model properly fumigated, and there has been no further sign of life. The model was cleaned carefully shortly after we acquired it, a break in the keel was repaired and a set of stern lanterns was made. Overall, the model is in spectacular original condition with all of its original paint, gilding and varnishing preserved. It is displayed on the original gilt metal supports.

CONSTRUCTION

Scale: 1/48 Hull length: 43in

This model is one of the last of the seventeenth-century-style

The quarter pieces take the form of putti holding nautical devices, an anchor on the starboard side and a cross-staff, shown here, on the port side.

models and was made just before radical changes in ship design severely reduced the carved decorative work. It is very much in the baroque tradition, adorned with numerous mythological creatures, hermaphrodites, dolphins, caryatids, nautical devices and symbols of sovereignty. The figurehead is a splendid British lion with a female companion designed to fit the awkward space between the lion's rump and the curve of the hair rail. As John Franklin was first to point out, the lion on this model may be the earliest example to sport a tail, because odd as it may seem, no earlier seventeenth-century British lion figurehead has one.[3] More interestingly, every subsequent eighteenth-century lion has a tail. This is quite remarkable when one considers that the models, as well as the ships themselves, were built in dockyards scattered throughout the British Isles, and for such a change to occur simultaneously everywhere suggests that it must have been specified in a Navy Board directive.

Other interesting decorative features include the belfry, which is carved with the figures of two sleeping cherubs, recumbent with their heads cushioned on their arms, and a miniature carved hourglass sitting between them. This strikes us as a wonderful vignette to serve as the canopy to the ship's bell, which after all, would ring to mark the change of watch and rouse sailors from their own slumbers. The hancing pieces that on most models are carved as crouching dogs, or stylised human figures, are unusually complex here and depict cherubs cavorting with dolphins. A typical feature of late seventeenth-century models is the profusion of carved, gilded heads wearing pointed caps that appear on nearly all belaying points, particularly on the beakhead and main bulkhead rail. The quarter pieces are cherubs holding anchors, and the taffrail proudly displays Queen Anne flanked by figures representing peace (holding an olive wreath) and plenty (bearing a cornucopia). Her motto, *Semper Eadem* (always the same), appears in a ribbon on the gallery rail below her bust. This model is one of only two or three to have an interesting arrangement of crossed vertical and horizontal stern timbers. The usual transoms have been fitted, but additional vertical timbers have been added, set into shallow mortices cut to receive them. These would have served to strengthen the stern.

One of the unique architectural features of this model is visible in the waist. There are a series of metal brackets that are hinged to the sides of the bulwarks in such a way that when fully extended, they are able to support a long plank running fore to aft between the fore deck and quarterdeck. Such a gangway allowing seamen to travel from the fore deck to the stern without needing to negotiate two sets of stairs is an innovative feature, and in this, the earliest example, the additional portable aspect lends flexibility that would not be present in a fixed version. Why this did not become standard practice after its appearance on this ship is a mystery, but this design feature was not to reappear for decades and was only much later to become universal practice.

Another fine feature is the grand bell staircase that appears on the quarterdeck, which would not look out of place in a stately country house. One of the more unusual constructional details on this model is the myriad of builder's notations in pencil that survive on the wooden parts, not only on internal raw wood surfaces, but also on finished external surfaces beneath layers of varnish. For example, the ribs and futtocks are numbered consecutively from the centrepiece, and are accompanied by the letter 'R' on the starboard side and 'L' on the port. This naval architect must have been a landlubber. Simon Stephens from the NMM visited once with his endoscope, and endoscopic examination of the interior of this model revealed additional pencil markings and a curious star-shaped stamp that marked the mid-section underneath one of the deck beams.

Literature

The following references include photographs and descriptions of this model:

Bowen, John, ed, *Model Shipwright*, 46 (December 1983), p. 2.

Brown, C, 'Down to the Sea in Ship Models', *Forbes*, 144, 11 (13 November 1989), pp. 336–40.

Clifford, Ann, 'The Elusive Admiralty Model', *Sea Heritage News* (September 1985), p. 4.

Franklin, John, *Navy Board Ship Models 1650–1750* (London: Conway Maritime Press Ltd, 1989), pp. 15, 127–31, 186, colour plates 9, 13.

Gardiner, Robert, ed, *The Line of Battle: the Sailing Warship 1650–1840* (London: Conway Maritime Press, 1992), p. 167.

Kriegstein, Arnold and Henry, 'The Kriegstein Collection of British Navy Board Ship Models', *Nautical Research Journal*, 38, 4 (December 1993), p. 221 and plate 7.

Lieberman, Cy and Pat, *The Mystique of Tall Ships* (Delaware, USA: Middle Atlantic Press, 1986), p. 129.

Parry, J E, et al, *Export of Works of Art 1983–84* (London: Her Majesty's Stationery Office, 1985), pp. 26–27, and plate 15.

Philbin, Tobias, & Endsor, Richard, *Warships for the King* (Florence, OR: SeaWatch Books LLC, 2012), p.54.

Winfield, Rif, *British Warships in the Age of Sail 1603–1714* (Barnsley: Seaforth Publishing, 2009), p. 76.

Exhibitions

Greenwich, England, National Maritime Museum, 1983–84.
Salem, Massachusetts, Peabody Essex Museum, September 2019–present.

Historical Perspective

THE *NORTHUMBERLAND* 3RD RATE OF 1702

In the years since the Committee on the Export of Works of Art decided to allow export of this model, we have often wondered whether we would ever be able to identify it. The concept that many models were proposals that were never built seems more and more unlikely, especially since we now know that at least some models were made while the ships were also under construction. A curious series of recent events have now allowed us to make a tentative identification. We can fix the date of the model rather narrowly to 1702. This is in part because the model has the exuberant quantity of carving characteristic of the seventeenth century, but a bust of Queen Anne at the stern. The model thus fits into the brief period

The form of the stern is continuous with that of the quarter galleries with the breast and counter rails wrapping around to become the quarter rails. The port quarter figure holds a cross-staff, and the starboard one carries an anchor. An interesting feature is the placement of vertical timbers between the transoms, which are found on a small number of models from the end of the seventeenth century.

after Anne assumed the throne in 1702 but prior to the restriction on carving instituted in June 1703 that put an end to the seventeenth-century decorative style. Moreover, the first owner, Thomas Herbert, 8th Earl of Pembroke, was Lord High Admiral for only a brief period from 1701 until May 1702, when Queen Anne appointed her husband to the post. Thomas Herbert must have acquired the model prior to May 1702 when he left office.

There were five 70-gun ships launched in 1701–02. Strict dating of the model to a ship of 1702 narrows the list of possibilities to only three: the *Restoration*, *Edgar* or *Northumberland*. We had no basis for choosing among these until Arnold made an interesting observation. There is a fine model of a 6th rate in the Thomson Collection at the Art Gallery of Ontario that so closely resembles our model in terms of the quality and style of carving, colouring, patination and paintwork, that we think it is very likely by the same maker. Even the gilt metal stanchions that support the keel are nearly identical to those on our model. The Thomson model has recently been identified as the *Nightingale* of 1702. The *Nightingale* was built by Robert Shortiss at Chatham Dockyard. Shortiss is known to have made models of his ships, as a model of the *Lizard*, a 6th rate built by Shortiss at Sheerness in 1697, is at the Pitt Rivers Museum in Oxford. This model has Shortiss's initials 'RS' carved in the cradles.

The lion figurehead on this model is one of the earliest to be shown sporting a tail. The scantling of each of the head timbers becomes progressively thinner toward the bow. On most models, the brackets of the head timbers are carved in the form of caryatids, and their gazes progressively shift from facing backward to facing forward, so that the foremost caryatid is peering over the rump of the lion, as on this example.

Interestingly, of the five 70-gun 3rd-rate ships that our model could represent, one, the *Northumberland*, was rebuilt by Robert Shortiss at Chatham in 1702. Based on this indirect evidence we are now willing to identify our model as the *Northumberland* of 1702.

The *Northumberland* was assigned to Sir George Rooke's fleet and participated in the attempt on Cadiz in August 1702, and saw action in Vigo Bay on 12 October, where she served as the flagship of Rear Admiral John Graydon. Sadly, the *Northumberland* had a short career, being wrecked on the Goodwin Sands during the 'Great Storm' of 1703 with the loss of her entire crew of 220. The wreck of the *Northumberland* was discovered by amateur divers in 1980 and is now designated a protected site under the Protection of Wrecks Act (1973).

THE LINE OF BATTLE INFLUENCES SHIP DESIGN

The reign of James II's daughter, Queen Mary, and her husband, William of Orange, ended the Catholic rule of James II but embroiled England in a long series of wars with France. On 30 June 1690, a combined Anglo-Dutch fleet faced a larger French one off Beachy Head. The allied commander, the Earl of Torrington, was cautious and defensive and lost the battle but preserved most of the English ships. This behaviour highlights a change in the way naval commanders used their fleets. Owing largely to the enormous time and resources that went into building and manning a fleet of ships of the line, tactics became largely defensive and daring exploits could not be risked. Merchant ships could no longer be recruited to fight alongside purpose-built men-of-war, and commanders would have to answer for any ships lost in battle. The line of battle became the safest, and therefore the standard, method of engagement. This imposed specific requirements for the performance of warships, for which extant designs were found wanting. Large three-deckers of the 1st and 2nd rates were extremely expensive. Two-deckers of the smaller sizes were either too lightly armed or would hog excessively and sail poorly if weighed down by upper deck guns carried on poop, quarterdeck and forecastle. The late seventeenth and early eighteenth century was a period of experimentation in naval design in a search for the ultimate ship of the line.

The result of all this innovation was to be the 74-gun 3rd-rate two-decker. The first of these supreme wooden warships would not be built for nearly fifty more years (see Chapter 14), but the *Northumberland* of 1702 represents an important early step in the right direction. Chief among her prescient features is the arrangement of her ordnance. Most of her guns are carried on the long main and upper gun decks, with no accommodation for forecastle guns, and with a crowded quarterdeck sporting seven guns per side. The short poop is bereft of arms, and this was an innovative feature in 1702, although it would become standard much later.

The taffrail is quite pacific for a warship, bearing two reclining female figures, one holding a sprig of palm leaves and one an overflowing cornucopia, representing victory and plenty respectively. The Tudor motto *semper eadem* (always the same) appears in low relief on the panel beneath the lights. There is an open stern gallery, and the upper counter serves as a decorated balcony railing. The delicate foliate carvings applied to the lower counter form an arch above the gun ports and terminate in dolphins.

CHAPTER 7

The *Marlborough*, 2nd rate of 1706

Acquisition

WE WERE VERY LUCKY when starting our collection that many beautiful and important models had slipped into obscurity over the years, enabling us to acquire them with little competition. Even when a rare and important Queen Anne three-decker such as the *Marlborough* was offered for sale in New York, there was little public understanding or appreciation of how exceptional an opportunity this represented. If this model had been sold in London, it may well have been retained as a national treasure. The *Marlborough* was the second of four Admiralty Board models offered in the Junius Morgan sale at Sotheby's in 1974. From the photographs we had acquired two weeks before the sale, it was apparent that this rigged ship model, unidentified at the time, was an early eighteenth-century dockyard model, notwithstanding that the auction catalogue dismissed it simply as a nineteenth-century copy. As recounted in Chapter 3, of the four models in the Morgan sale, we only bought one at the auction, the *Coronation*, which in our opinion was the rarest model on offer that day. This one was, however, a close second. In this case, our advantage of knowing it was an authentic Admiralty Board ship model was significantly undermined by Sotheby's announcement just before the models were offered that the vendor, John P Morgan II, had discovered correspondence indicating that the first model to be sold was probably of the *Coronation*, and the second model was the *Marlborough*, and that both had been on view at the London Museum in England for many years. A photograph showing the *Marlborough* in the museum was produced and both the letter and photograph were to be included in the lot. Although we had not made plans to bid, when the activity paused at a surprisingly modest level, Arnold jumped in with a bid of his own, but the bidding evolved into a contest between a fellow standing at the back of the tent and a gentleman in the front row. When the dust settled, the winner was the gentleman in the back and the price was higher than that for the *Coronation*. The third model went to the same bidder, and the fourth model did not sell that day at all. Eventually, all four models joined our collection. One we acquired from Sotheby's privately just one week after the sale, but eleven years were to pass before we acquired the final model, as detailed in Chapter 4.

The gentleman at the back of the tent turned out to be Mr Landrigan, of Landrigan and Stair, who were just establishing an antique furniture shop on Madison Avenue, New York. Over the next few weeks they brought photographs of their model to the NMM,

Left: The *Marlborough* nicely embodies the elegance and grace that typifies the Queen Anne period and finds expression in the appearance of her navy as well as in the style of furniture and architecture. The profusion of gilded baroque carving of the seventeenth century has been severely restrained and colouring minimised through the use of varnished wood surfaces. The occasional moulding, rail or strake is picked out in black or red. The spritsail topmast, as shown here, persisted well into the eighteenth century on three-deckers, but the jib-boom had already replaced the spritsail topmast on smaller ships of this date and disappeared entirely by around 1740.

Above: This old photograph shows the model as it appeared in the London Museum prior to re-rigging in 1920. At that time it was ludicrously rigged in nineteenth-century style with a bowsprit, jib-boom and mizzen gaff. R C Anderson rigged it correctly at his own expense for the benefit of the public, but not long after he had finished, the model was sold, eventually entering Junius Morgan's private collection in America.

An unusual 'single' equestrian figurehead graces the bow. Roundhouses are apparent on the beakhead bulkhead, as well as seats of ease on the head gratings. The cathead supporter merges with the middle headrail and runs across the foremost middle gun deck port. Rigols keep rainwater from running into the gun ports. A sideways-facing flight of stairs leads from the foredeck to the upper gun deck entered by a companionway located just before the foremast. The galley flue is visible, topped by a cowling fitted with handles, so it can be rotated to face downwind.

The quarter figure, carved from a single piece, represents Fame, who with trumpets sounding, stands on a docile-looking lion borrowed from the Churchill coat of arms. The crest of the Churchill arms features a lion bearing a standard displaying the symbol of an open palm, as depicted here.

where the authorities confirmed that the model was indeed a rare and important dockyard model of the *Marlborough*, built in 1706. We began a protracted negotiation to purchase the model, but the dealers intended to use it as a centrepiece of their booth at the Winter Antique show held at the Armory in New York, and they were reluctant to sell it before the show. The price at the show was too high for us, and we waited anxiously hoping that the model would remain unsold. Arnold saw the model on opening day, sitting proudly front and centre in the Landrigan and Stair booth and looking every bit the finest and most important item in the entire show. When Arnold returned on the third day of the ten-day show, the model was gone. He presumed it had been sold, and taking a long-term view, asked for details concerning the identity of the buyer. To his surprise, he learned that it had not been sold yet but had been hurriedly loaned to a potential institutional buyer. This was actually encouraging news. If the customer decided against purchase, the model would still be available, but would be off view for several days during the show. The institutional board was unable to reach a consensus about purchase, and the model remained unsold. We finally bought the *Marlborough* several months later, just as Landrigan and Stair were preparing to ship it to a show in Texas. We did not want the anxiety of waiting to see what might happen in Texas and were also concerned about possible damage during shipping. Models of three-deckers, the

Despite the small scale, there are no fewer than five carved figures on the taffrail including Queen Anne, in the centre, being presented a wreath and her royal regalia by flanking male and female figures aided by a winged cherub to port and a young warrior to starboard. The latter are displaying armorial shields. The gallery rails are finely wrought of brass sheet fretted to form stylistic panels between finial-capped stanchions. The centrepiece of the upper gallery incorporates the cypher of Queen Anne (AR), while that of John Churchill, Duke of Marlborough (JCM) is incised in the lower one.

largest ships in the fleet, are extremely rare in private hands, and we had already come too close to losing the opportunity to acquire this example. In our experience, collectors regret the objects they miss, not the expense of those they acquire.

Provenance

This model and the *Coronation* were both once the property of John Vaughn, Earl of Carbery, and a Lord of the Admiralty during 1683–84. He must have acquired the *Marlborough* model sometime between 1706 and his death in 1713. When Sir Richard Gough Kent bought the Carbery estate, subsequently known as Gough House, he presumably also bought the contents, as both models remained at Gough House, Chelsea, until 1911. In that year they were lent by Mrs Anstruther Gough Calthorpe to the London Museum, which was housed in Kensington Palace. While on public view, the models were seen by R C Anderson, who wrote a letter to the owner to express his opinion that the rigging of the models, of nineteenth-century vintage, was erroneous, and therefore the display of these models perpetrated a hoax on the public. Anderson was the acknowledged expert on rigging of this period, having written the authoritative book *Rigging in the Days of the Spritsail Topmast*.[1] The outcome was that Anderson was given the models for re-rigging, which was carried out over the course of one year or so, in 1920, at his own expense. Most of the work was actually done by Anderson's assistant, L A Pritchard, under Anderson's supervision, and when the *Marlborough* went back on display it had rigging accurate for the period. The public, however, did not benefit for long, since in February 1924 Mrs Calthorpe sold her models to the antique dealer, Rochelle Thomas. Despite his assurance to her that they would not be exported, he sold them to the New York dealer Max Williams, who shipped them to his Madison Avenue gallery. They were bought by the collector Junius S Morgan Jr, grandson of the financier J P, in whose residence in Glen Cove, Long Island, they remained until sold by Sotheby's in 1974.

Description

CONDITION

This model is in excellent, original condition. When acquired, one of the stern lanterns was missing, and a replacement was made by Philip Wride. As mentioned above, the model was re-rigged in 1920 by R C Anderson and L A Pritchard while on loan to the London Museum.

CONSTRUCTION

Scale: 1/72   Hull length: 31in

This is the first of three models in our collection built to this small scale, which enjoyed some popularity in the early eighteenth century. It is unusual to find an equestrian figurehead on any ship other than the largest 1st rates. The presence of an equestrian figurehead on a 2nd rate is therefore a deviation from usual

The foredeck bulkhead is painted black with yellow trim, and the belfry canopy takes the form of a plain arch. A bench is situated at the foot of the belfry, incorporating small lights to help illuminate the middle gun deck. A pissdale is visible just abaft the twist gangway stair, fitted with a drainpipe. The jeer capstan is of the drumhead type and equipped with six whelps and ten bars.

practice, though a painting by Peter Monamy of the *Blenheim* built in 1709, a 'sister' ship to the *Marlborough*, shows that it also sported a similar rearing horse. The rider on the *Marlborough* figurehead, depicted as a classical Roman warrior brandishing a sword over his head and holding a shield decorated with an effigy of Britannia enthroned, can be identified as John Churchill, the 1st Duke of *Marlborough*, by the cypher containing his elegantly carved initials on the saddle, just behind the rider's right and left stirrups. We know the identity of the hapless fellow being trampled beneath the horse's hooves from a first-hand account by Marcus Luttrell, a witness to the launching of the ship. Luttrell recorded in his memoirs that the *Marlborough* was decorated with an effigy of the Duke trampling over Marshall Tallard, his archrival on the battlefield of Blenheim in 1704. On the model, as presumably on the ship itself, the luckless Tallard is victimised twice, both starboard and port.

Atop the quarter galleries on either side there is a reclining figure of a classical warrior resting on a shield that bears the AR cipher and holding a Tudor rose in his hand. Aside from the figurehead, the decoration of the stern provides additional evidence of the identity of the model. The Churchill arms appear in minute detail in a finely carved escutcheon just under the lower row of lights (windows) at the stern. The lions squatting at the base of both quarter galleries are shown holding standards with flags that are decorated with open palms, the same symbols that can be seen emblazoned on the gates to Blenheim Palace, the Churchill estate in Woodstock, England. There is also an interesting comment on the relationship of Churchill to Queen Anne in the beautifully rendered metalwork balustrades appearing at the stern balconies. The upper galleries are decorated with elegantly entwined 'AR' ciphers of Queen Anne, shown dexter and in mirror-image to sinister, while the lower-tier balustrades are decorated with the elegant ciphers 'JCM', also shown in mirror image, representing the initials of John Churchill, the Duke of Marlborough. In 1706 Churchill had just achieved a resounding victory on the battlefields of France and was roundly hailed as a war hero. Both he and his wife were favourites of the Queen, a relationship that would sour over the ensuing years, but was in full bloom when the *Marlborough* was being built. The metalwork on the railing is unique among models of this period and is all the more remarkable given that it was expertly cut and filed by hand.

On wooden ships of this period, the hull planking was secured to the underlying hull timbers by large wooden nails, dowels really, called appropriately enough 'treenails' or more commonly by the slightly abbreviated form, 'trenails'. On most models the planking is fastened to the hull by brass pins, often placed in a seemingly random distribution but where the positions of at least some are probably dictated by the curves and bends of the hull. Occasionally, and often on the best-crafted examples, the planks are held in place by miniature trenails aligned in neat rows as in the full-size ships, and this is the case in the *Marlborough*. The heads of the trenails can be seen as regularly spaced small round spots on the surface of the planks. By shining a light inside the hull, it is possible to observe that the trenails consist of slightly tapering miniature dowels that were driven into holes drilled through the planks and hull frames and then planed off flush on the exterior, but not trimmed on the interior, so that rows of protruding wooden pins can be seen neatly spaced and all of equal length extending throughout the interior of the hull. The appearance suggests a porcupine turned inside out.

Literature

The following references include photographs and descriptions of this model:

Anderson, R C, *Catalogue of Ship Models* (London: Her Majesty's Stationery Office, 1952), p. 12.

Anderson, R C, 'Comparative Naval Architecture, 1670–1720', *Mariner's Mirror*, 7 (1921), pp. 308–15.

Brown, C, 'Down to the Sea in Ship Models,' *Forbes*, 144, 11 (13 November 1989), pp. 336–40.

Edelstein, A E, *Art at Auction 1973–74* (New York: The Viking Press, 1974), p. 242.

Franklin, John, *Navy Board Ship Models 1650–1750* (London: Conway Maritime Press Ltd, 1989), p. 29, 140–42.

Kobak, Laurence B, 'British Admiralty Model Collection', *Sea Heritage News*, 4, 12 (1983), p. 6.

Kriegstein, Arnold and Henry, 'The Kriegstein Ship Model Collection', *Nautical Research Journal*, 27, 2 (June 1981), pp. 90–1.

Kriegstein, Arnold and Henry, 'The Kriegstein Collection of British Navy Board Ship Models', *Nautical Research Journal*, 38, 4 (December 1993), p. 223 and plates 11, 12.

Laughton, L G C, *Old Ship Figureheads and Sterns* (London: Halton & Truscott Smith Ltd, 1925), p. 76, 204.

Lavery, Brian, *The Ship of the Line* (Annapolis, Naval Institute Press, 2014), p. 44.

Liberman, Cy and Pat, *The Mystique of Tall Ships*, 132 (Delaware, USA: Middle Atlantic Press, 1986).

Nance, R Morton, *Sailing-Ship Models* (New York and London: Halton & Truscott Smith Ltd, 1924), p. 73, plate 62.

Nance, R Morton, *Sailing-Ship Models* (New York and London: Halton & Truscott Smith Ltd, 1949), p. 58, plate 55.

Pritchard, L A, 'Model of 'Victory' in London Museum', *Mariner's Mirror*, 9 (1923), p. 157.

W S, 'A Register of Models', *Mariner's Mirror*, 3, 2 (February 1913), pp. 57–8.

Winfield, Rif, *British Warships in the Age of Sail 1603–1714* (Barnsley: Seaforth Publishing, 2009), p. 38.

Exhibitions

London, Kensington Palace, London Museum, 1911–24.

The allegorical equestrian figurehead depicts John Churchill, the Duke of Marlborough, as a classical warrior trampling over Marshall Tallard, who was the French commander at the Battle of Blenheim. Churchill was victorious at this decisive battle of the War of the Spanish Succession, and this ship as well as another 2nd rate named the *Blenheim*, commemorated his achievement. The Duke's initials, 'JCM', are carved into the saddle, and the hapless Tallard also appears on the port side, where a carving of Britannia seated decorates the Duke's shield.

Historical Perspective

The *Marlborough* was built at the Blackwall dockyard by William Johnson and launched in 1706 during the early years of the reign of Queen Anne. It was actually a rebuild of the seventeenth-century 2nd-rate ship, the *St Michael*, meaning that the hull of the older ship was used in building the new one as an economy measure. The *Marlborough* was rebuilt again in 1725, and following an illustrious career, she sank in heavy weather on 29 November 1762. A very fine dockyard model of the *St Michael* also survives in the collection of the National Maritime Museum in Greenwich, England, and this allows a comparison of the two ships. Both are 90-gun warships, but the similarity essentially ends there. The dimensions and shape of the hulls, the gun port arrangements, the angles of the decks, the bulkhead placements, and, of course, the decorations of the two ships bear no resemblance to each other. This was clearly a major rebuild, meaning some of the timbers of the older ship were probably recycled and used in building the hull of the new one. There would have been very little, if any, savings, but the Navy Board was always looking for ways to economise.

CARVED AND GILDED DECORATIONS DISAPPEAR ON MODELS AND SHIPS

Despite the enormous cost of shipbuilding, or perhaps because of it, the sailors themselves often went without pay. Some Englishmen, angered over years of unpaid service, shifted allegiance to the Dutch during the wars. As noted earlier, it was a disgruntled English pilot on board a Dutch ship who helped guide De Ruyter's invasion fleet up the Medway. Financial constraints helped to stimulate the particularly interesting transitional period in naval architecture that occurred during the early years of Queen Anne's reign. The sweeping curves and exuberant baroque decoration of the seventeenth century were rapidly superseded by cleaner lines and more economical decoration.

Shipbuilding was the grandest and costliest enterprise in England when Queen Anne came to the throne in 1702, and she took advantage of a relatively quiet period of naval inactivity to attempt cost-cutting measures. Interestingly, these consisted of reducing the money spent on decorations. The gorgeous baroque decorations of seventeenth-century British warships combined with the mighty force of the gun batteries personified the glories of the kingdom itself, or at least the wishful impression of glory and power that the monarch sought to project. As an instrument of diplomacy, a richly adorned British man-of-war served as a splendid ambassador in foreign ports and helped establish the relatively small island country as a world-class force. But splendid and impressive as these ships were, the costly decorations were peripheral to the main function of a warship, which was to serve as a portable gun battery. Shortly after Queen Anne assumed the throne, in June 1703, an Admiralty order was issued specifying that '… the carved work be reduced only to a lion and trailboard for the head, with mouldings instead of brackets placed against the timbers; that the stern have only a taffrail and two quarter pieces, and in lieu of brackets between the lights of the stern galleries and bulkheads, to have mouldings fixed against the timbers'.[2] The term 'brackets' refers to the boldly carved vertical timbers usually in the form of a human figure with female torso and breasts, but often with a bearded male head. These carved brackets were gilded and placed against nearly every vertical timber on ship's bulkheads in the seventeenth century, but they disappeared completely in the early eighteenth century. Only in the lineage of royal yachts did the sumptuous gilded carved decoration persist, and the carved work adorning the *Royal Sovereign* yacht of 1804 would not have looked out of place on a ship built 100 years earlier.

These changes in warship design and decoration can be glimpsed in paintings and drawings, but since none of the ships themselves have survived, they are best documented in those few surviving three-dimensional models of the period such as this one. For example, the carved and gilded wreaths surrounding nearly all the upper gun port lids on seventeenth-century ships, were, as on the *Marlborough*, reduced to simple circular mouldings on the quarterdeck ports only. Similarly, as specified by the Admiralty, the elaborate carved brackets that adorned the bulkheads of nearly all seventeenth-century ships have been replaced here by simple mouldings. Although elaborate carvings were confined mainly to the figurehead and stern, the quality of the carving is superb and probably more apparent on models such as this one because the carvings at this period were simply varnished, while in the seventeenth century the fine detail was obscured by coats of gesso and gilt. For just a few years at the beginning of the eighteenth century, not only the carving, but also the extensive gilded and painted decoration typical of most Admiralty models was severely reduced. The models produced over the years 1703–10 were unique in being finished simply by applying varnish to bare wood. The result was that the models, generally made of fruitwood such as box

THE *MARLBOROUGH*, 2ND RATE OF 1706

and pear, resemble wooden sculpture more than at any other period, and permit a closer glimpse of the woodcarver's art than may be apparent in the gilded and painted versions of earlier and later years. We have two examples of models from this period, the *Marlborough* and the *Diamond* model. Both of these models are entirely complete and original, and the hand of time has mellowed the varnished surfaces, darkening the wood and bestowing a very pleasing warm glow.

Compared to earlier three-deckers, the sheer is noticeably reduced, and the pronounced tumblehome creates the impression of a very seaworthy vessel. From this perspective, it is not difficult to imagine how the ship would have looked at sea.

Models built at the beginning of the eighteenth century, such as this one, are particularly interesting because of the transitional features they display as ships were rapidly becoming more modern and efficient. Carved gun port wreaths and cathead brackets are making nearly their last appearance here, while the pronounced sheer of the previous century, along with the characteristic high stern, has essentially disappeared.

CHAPTER 8

The *Diamond*, 4th rate of 1708

Acquisition

WE ARE LUCKY THAT a fad for ship models in the 1920s spawned a series of picture books that inspired us and set us on the trail of many of the models that now appear in this book. We first learned about the existence of this model when we came across a copy of a 1924 book entitled *Sailing-Ship Models* by R Morton Nance. The model appears in two illustrations, where it is ascribed to the collection of Colonel Rogers.[1] We saw this book in 1974 and already knew that upon his death, Col Rogers had bequeathed his collection to the United States Naval Academy, and that it was currently housed at the Naval Academy Museum in Annapolis. A quick check of the catalogue of this collection, however, did not show any evidence of this particular model. A call to the curator confirmed that this model had not been in the collection when it was gifted, and that its whereabouts were unknown. Our next step was to contact the heirs of Col Rogers, and a meeting with family members taught us that the model had not been left behind in the house, but rather, had left the collection prior to his death, most likely in a divorce settlement.

For a number of years no further progress was made. Then, in 1979, in a conversation with the curator at Annapolis, we were told that this model had reappeared. It was now on loan to the Mariners' Museum in Newport News, Virginia! Henry wasted no time in travelling to the museum to examine the model, which was still beautifully preserved. The curator at the museum explained that they were in discussions with the owner and hoped to buy the model for their permanent collection. We agreed to forbear contacting the owner until the museum had concluded their negotiations, despite the fact that the label on the model provided the owner's name. We restrained ourselves for a full year, and then contacted the museum to learn to our delight that the purchase process had stalled because they were unable to agree on a price. We felt the time had come to proceed on our own, and we set out to locate the owner ourselves.

This was in the time before the World Wide Web, and despite knowing his name, our search for his whereabouts was to prove more challenging than we had expected. First, there was no listing for anyone by this name in any of the telephone directories for areas within three hours' drive of the Mariners' Museum. Henry finally decided to try a less conventional gambit and called the Motor Vehicle Registry in Virginia and asked the polite woman who answered if a gentleman by this name had a valid driver's licence in the state. To our surprise, she responded with, 'Is that the gentleman on such-and-such street?' 'Yes!' Henry said and hung up. We already knew that there was no phone for him at this address and we needed to be creative again. Henry called the Post Office for this small town and asked to speak with the man who delivered mail to the address we had been given. When Henry asked the postman if he knew this man, he said yes, but that he had moved over a year ago and was now living in a small town in New York. The elusive

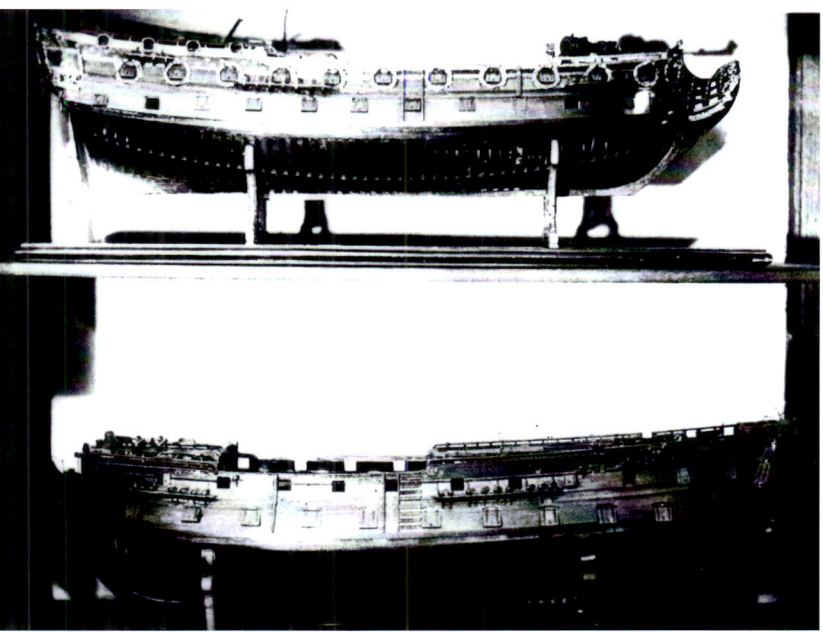

A photograph showing the *Diamond* (below) on display at Cuckfield Park, where it remained for 200 years. The *Diamond* was part of the collection of models formed by Charles Sergison, who was Clerk of the Acts of the Admiralty during the reign of Queen Anne.

The colour scheme of this model is simply varnished wood with black and red trim. Many of the black mouldings and strakes are made of ebony. The absence of gilding and decoration is typical of models built during the Admiralty of Queen Anne.

owner was listed in the phone book at this address, and in two minutes we were talking to him! Not long after, on 14 March 1981, the model was in our collection.

Provenance

The first owner of this model was Charles Sergison, Clerk of the Acts to the Admiralty Board 1689 to 1719. Upon his death it remained in a special room on his estate, Cuckfield Park, in Surrey, England, along with other models in his collection until they were brought to the attention of R C Anderson, who was given the opportunity to study and repair them. In the midst of his efforts, in 1922, the entire collection was sold to the American collector Colonel Henry Huddleston Rogers. This caused considerable consternation among British naval scholars and historians, but other than editorial complaints in several journals, nothing could

Right: The fine quality of the carvings, as for example the lion and his little female companion seen here, can probably be best appreciated on Queen Anne models when they were simply finished with a coat of varnish. Two seats of ease are placed just aft of the figurehead, with another pair tucked in between the head rails and the roundhouses. The berthing rail, which was used to support a tarpaulin, not to provide a handhold, consists of a rope suspended between ringbolts and supported by a pair of stanchions. The brackets flanking the gammoning slot are unusual features provided to guide the gammoning ropes. The main head rails terminate in unusual volute scrolls.

be done to prevent their export. Once in America, Sergison's unique collection of early Admiralty models was lent by Col Rogers to the Metropolitan Museum of Art in New York, where this model was on display in 1922–32. Sometime between then and Rogers' death in 1935 this lovely 4th rate left Col Rogers' collection. It was loaned to the Mariners' Museum in Newport News, Virginia, in 1977–80 and was bought by us in 1981.

Description

CONDITION

This model remained in its original walnut veneer case until the third decade of the twentieth century and is remarkably well preserved. There is no restoration on the model. We had it carefully cleaned after purchasing it, and reproduction cradles were fitted based upon photographs of the originals.

CONSTRUCTION

Scale: 1/48 Hull length: 35in

The model is made almost entirely of pear wood with ebony trim and inlay and restrained colouring, limited mostly to red around the gun ports and on the inside of the port lids. The carvings are varnished rather than gilded, and this permits a full appreciation of the skill of the model maker. The figurehead is a magnificent lion

THE *DIAMOND*, 4TH RATE OF 1708

The stern contains a single deep gallery with a pair of central double-hung doors to the captain's cabin. The hexagonal stern lanterns are glazed with small glass panes and equipped with small clusters of candle holders. The 'peace at home, war abroad' gun port configuration means that the aftmost four upper deck ports are fitted with glazed lights. A small aperture is provided in each light, presumably for small arms fire. A pair of glazed scuttles are visible, fitted with shell-form canopies. The lower finishing of the quarter gallery takes the form of a winged merchild; a similar carving is found on the trailboard. The upper finishing depicts a crowned rose and thistle, a badge adopted by Queen Anne following the Act of Union with Scotland in 1707. The quarterdeck gun port wreaths echo this theme; one sports roses, one thistles.

with a hint of the Asian influence in decoration that was in fashion at the time as reflected in the robust curling mane and tail. An old-fashioned feature for this period is the little female companion carved to fit the awkward space between the beast and the hair rail. An interesting and unusual feature is the finely fashioned charming 'merchild' carved in relief along the trailboard. This carving depicts a youth with wings and a twisted fish tail and is carved in such a way as to be visible from both port and starboard. The merchild makes his appearance again at the stern, where he forms the lower finishing of the quarter galleries. Perhaps a reader will know what this figure signifies, but his meaning evades us. Centred on the taffrail is a beautiful carving of a bust of the Queen, round-cheeked and wearing a strand of pearls and surrounded by a wreath of ivy. Suspended between winged cherub heads below is a banner with the legend: *Semper Eadem* (always the same), the Tudor motto used by Queen Anne. Flanking the Queen on her right is the Greek god Apollo, ivy wreathed and recognisable by the lyre cradled on his knee, and to the left is the war-like god Ares, attired in armour and bearing a sword and shield. A nice touch is the carved Britannic cross on the face of Ares' shield. The quarter pieces are images of Athena or Britannia herself, bearing a shield and spear. The stern lanterns are hexagonal with the tapered shape typical of the period, but are of exceptional quality, and by peering through the mica panes, one can see clusters of candle holders carefully fitted inside.

The gun ports are arranged in the 'peace at home, war abroad' configuration, meaning that the aftermost four upper deck gun ports are fitted with glazed windows, a great convenience for the officers housed in the aft cabins, and there are additional glazed scuttles between the ports fitted with shell-form canopies supported on brackets. The stern cabins are finely detailed and include mouldings, panelling and ebony inlay. The robustly carved baroque hancing pieces of former decades have gone out of fashion, and instead the upper rail mouldings terminate in fine volute scrolls providing an elegant and refined look. An unusual feature is the provision of stanchions rigged with rope for stowing hammocks on

The taffrail is beautifully carved, centred by a portrait bust of the Queen and a ribbon bearing her motto, *semper eadem*, suspended between two cherub heads. Apollo, wreathed in laurel and holding his lyre, is seated to the Queen's right, while Ares, brandishing a sword and buckler, is seated on the Queen's left. The quarter pieces are probably figures of Athena holding a spear and hoplon shield, but the British Union Jack decorating the shield also suggests Britannia.

the quarterdeck rail, and stanchions serving as guard rails on the fore deck. Centred on the forecastle bulkhead is a bench with a brass grille below, possibly to house poultry. Inside the belfry there is a nicely shaped wooden bell. The needs of the officers and sailors have not been overlooked, and there are four seats of ease provided at the beakhead and pissdales fitted in the waist.

The model demonstrates the effect of the order of 1703 restricting lavish decoration. Gone are the gilded carvings characteristic of seventeenth-century ships; so too are the figurative caryatids and brackets that decorated the head and bulkheads. This model and the ship it represents belong to that brief moment in time when shipwrights had not fully abandoned the old ways, but had not yet fully embraced the future. Beautifully carved wreaths surround the three poop deck gun ports, and the cathead brackets are carved with the crouched bearded figures characteristic of seventeenth-century practice, but their appearance on this model may mark the last instance that these decorative elements graced a British warship. In contrast to these old-fashioned elements, the model has starkly modern lines, with almost no sheer, and displays what may be one of the earliest examples of a steering wheel.

Literature

The following references include photographs and descriptions of this model:

Anderson, R C, 'The Cuckfield M.S.S. and Models', *The Mariner's Mirror*, 5 (1919), p. 154.

Bowen, Frank, *From Carrack to Clipper* (London: Halton & Truscott Smith Ltd, 1927), p. 33, 36, plates 32, 33.

Bowen, Frank. *From Carrack to Clipper*, 2nd ed, (London: Halton & Company Ltd, 1948), p. 34, plate 32.

Chatterton, E, Keble, *Ship Models* (London: The Studio Ltd, 1923), p. 22.

Franklin, John, *Navy Board Ship Models 1650–1750* (London: Conway Maritime Press Ltd, 1989), pp. 23, 33, 34, 57, 147–49, colour plate 10.

Gardiner, Robert, ed, *The Line of Battle: The Sailing Warship 1650–1840* (London: Conway Maritime Press, 1992), p. 167.

Kriegstein, Arnold and Henry, 'The Kriegstein Collection of British Navy Board Ship Models', *Nautical Research Journal*, 38, 4 (December 1993), pp. 222–23, plate 10.

Nance, Morton, *Sailing-Ship Models* (London: Halton & Truscott Smith Ltd, 1924), pp. 73–74, plates 64 and 65.

Nance, Morton, *Sailing-Ship Models*, 2d ed, (New York: Halton, 1949), pp. 57–8, plates 49 and 59.

Winfield, Rif, *British Warships in the Age of Sail 1603–1714* (Barnsley: Seaforth Publishing, 2009), p. 168.

Exhibitions

New York, New York, The Metropolitan Museum of Art, 1922–32.
Newport News, Virginia, The Mariners' Museum, 1977–80.

Historical Perspective

This model represents a 54-gun 4th rate, and while no date or name appears anywhere on the model, there is good reason to believe that it represents the *Diamond* of 1708. The model can be dated to 1707 or 1708 on the basis of evidence contained in the carved decoration. The persistence of seventeenth-century decorative elements suggests that this model belongs to the early period of Queen Anne's reign. However, above the quarter galleries there is a carving of a rose and thistle conjoined and surmounted by a crown. This is the badge that Queen Anne adopted on the occasion of the Act of Union of Scotland with England that occurred in 1707. The model cannot, therefore, date from earlier than 1707. The ship represented by the model is one of a small class of early eighteenth-century 4th rates. The measurements of the gun deck length and breadth of the model translate to 115 ft and 32 ft 6 in, respectively. These measurements are a good match for one of the 4th rates named the *Diamond* that was built in 1708. The model has distinctive diamond-shaped decorations on the upper counter of the stern and on the stools of the quarter galleries. It is likely that these are emblematic of the name. The *Diamond* was built by Johnson and launched at Blackwall in 1708. She was broken up in 1721.

This model is decorated to commemorate the Treaty of Union between England and Scotland that took effect on 1 May 1707. Under the terms of this agreement, Scotland lost its parliament and independence and could ostensibly no longer threaten the Protestant English monarchy. This subjugation was a clear triumph of English diplomacy, accomplished as it was without bloodshed and in the face of overwhelming opposition from the Scottish populace. It succeeded largely because Scotland was

THE *DIAMOND*, 4TH RATE OF 1708

The forecastle bulkhead is recessed below the foredeck. A bench seat with high shaped supports and a grating is centred on the bulkhead below the belfry. The belfry canopy takes the form of a simple arch, with a suspended wooden bell. A pissdale can be seen against the starboard bulwark. Arched rigols are fitted above those gun ports that are not sheltered by channels.

bankrupt and not in a position to refuse the £398,000 that England was willing to pay in compensation to Scotsmen at all levels of society. This compensation was for the loss of half of the wealth of the country, which had been squandered on an ill-conceived scheme to establish an overseas Scottish colony in Central America. This enterprise is known for the name of the Panamanian isthmus where the settlement was to be established, Darien.

FAILURE OF THE DARIEN SCHEME PROMOTES THE UNION OF ENGLAND AND SCOTLAND

The Anglo-Dutch wars of the second half of the seventeenth century were particularly hard on the Scots. The English Navigation Act, which was the stimulus for the first of the three wars with Holland, was aimed at giving England some of the lucrative carrying trade that the Dutch enjoyed. It was, however, applied against Scotland as well, and coupled with the disruption of war and years of famine, helped to place Scotland in a precarious socio-economic condition by 1695. Many in Scotland dreamt of imitating Holland and growing rich through overseas trade. Such a mix of hope and desperation can fuel a gambling impulse. It was in this context that William Paterson, a financial adventurer, hatched a scheme for erecting a Scottish East India Company, and from this idea, 'in great haste and excitement, was drafted one of the most noble, vainglorious and disastrous Acts ever passed by the Parliament of Scotland'.[2]

In June 1695, the 'Company of Scotland Trading to Africa and the Indies' was founded. The premise of this enterprise was an audacious scheme to establish a colony on Darien, located on the isthmus of Panama. From this strategic position, it was argued, the Company could ferry goods overland from the Pacific to the Atlantic. This would provide a safer, quicker and more reliable alternative to the hazardous sea voyage around the South American continent to the Far East, and, indeed, would also serve as a shorter alternative to the Dutch route around Africa.

As a young man, Paterson spent seven or eight years in the West Indies, where he first heard of Darien, 'the green and beautiful country on the northern coast of Panama, where the earth yielded fruit without cultivation, where noble, naked Indians knew the secrets of un-mined gold, and where lush mountain valleys led to the Pacific sea',[3] or so he was told. William Paterson had not set foot upon any part of the Central American mainland, but this false image captivated him and was destined to ruin his life along with that of many others and alter the destiny of a nation.

Paterson returned to Europe and began to hawk his dream of colonisation, initially without success. After receiving no encouragement in Amsterdam or Berlin, he settled in London and became a successful merchant. In 1693 he successfully represented a group of colleagues who proposed a scheme for obtaining credit upon Parliamentary security, resulting in the formation of the Bank of England the following year. Paterson was one of its first directors, but he did not get along with his associates and in a demonstration of characteristic poor judgment, he resigned in 1695, in time to avoid any significant financial benefit.

It was at this same time that mercantile interests in his homeland began to look for a way to profit from colonisation in Africa or America. These merchants appealed to the Scottish Parliament, which enthusiastically embraced Paterson's plan to break the monopoly of the British East India Company. So it was that the 'Company of Scotland Trading to Africa and the Indies' was formed. A trading colony was to be established in Darien, and the company would control trade across the Pacific. The Spanish, based in nearby Cartagena, claimed control of Panama but this detail did not deter the Scots directors of the company.

The English East India Company felt threatened by this potential rival Scottish company, and persuaded William of Orange to ban English investment. This did not prove an impediment, as enthusiasm in Scotland was unbridled and thousands, both rich and poor, rushed to subscribe. In a matter of weeks £400,000, half the total capital available in Scotland, had been raised. This support was largely founded upon a report of conditions in Darien written by Lionel Wafer, a buccaneer and surgeon whose journal was brought to the company's attention by Paterson. Wafer had spent several months on the isthmus of Darien, had lived with the natives there and written a most complimentary narrative. He lauded the natural beauty and richness of the land, attested to the friendliness of the inhabitants, and described an Eden of boundless resources. Scots who read this account were somehow inclined to disregard both the presence and experience of the Spanish in the region. Spain, having originally colonised this part of Central America two centuries earlier, had rejected the swamps of Darien in favour of Portobello, Cartagena, and Panama, where sizeable garrisons were stationed.

On 4 July 1698, an expedition of five ships left Leith harbour bound for Darien, carrying 1,200 settlers, among them William Paterson. They also carried 4,000 wigs, 1,500 Bibles and tens of thousands of combs. They reached Darien on 2 November, seventy people having died on the voyage. They christened the peninsula New Caledonia, and through the winter they succeeded in building a stockade and a ragged village of huts. The rain-soaked land resisted attempts at cultivation, and the natives had no interest in the trinkets they were offered. Spring brought even more rain, as well as sickness, and men were dying at a rate of ten or twelve a

day. The plight of the settlers was worsened by the ban on trade imposed against them by the English, and finally a threatened attack by Spanish forces led to abandonment of the settlement. On 18 June, all of the surviving colonists set sail for Scotland save for six, who preferred to remain and die on land. In the end, only one ship reached Scotland, with fewer than 300 survivors.

Unfortunately, they arrived too late to prevent a second expedition from leaving on 18 August 1699. This small fleet carried 1,302 men and women aboard three ships, and 160 perished before even reaching the abandoned colony, which had been burned by the Spanish. They attempted to rebuild, but were faced with the same inhospitable conditions to which their predecessors had succumbed. Finally, in March 1700 the remaining colonists surrendered to a besieging Spanish force, who allowed them to depart on their remaining ships. Francis Borland, a minister who survived to reach Scotland, recorded the misery of the first leg of that return voyage:

The upper deck is constructed with two tiers of short carlings each side fitted with ledges, while the foredeck, quarterdeck and poop deck are more lightly constructed, with deck beams only. A stern cabin fits snugly into the aft quarterdeck. A central gangway extends forward to the mainmast, and quarterdeck gangways extend an equal distance, terminating in stairs to the maindeck.

> Malignant fevers and fluxes were the most common diseases, which swept away great numbers from among us. From aboard one ship, the Rising Sun, they would sometimes bury in the sea eight in one morning, besides what died out of the other ships. And when men were taken with these diseases, they would sometimes die like men distracted, in a very sad and fearful-like manner; but this was yet more lamentable to be seen among these poor, afflicted and plagued people, that for all God so afflicted them, yet they sinned still the more, were as hard and as impenitent as before, would still curse and swear when God's hand was heavy on them, and their neighbours dying and dead about them.[4]

Only a handful saw Scotland again.

It was against the background of this failed enterprise and associated impoverishment, that the English offer of £400,000 was accepted in exchange for Union. Jacobite rebellions continued for a while, but were no more successful than the Darien scheme. A new flag was created, combining the red cross of St George with the blue cross of St Andrew, giving birth to the Union Jack that would identify British warships for the rest of the century.

Above: Hammock cranes, rarely fitted, have their earliest appearance on this model and are fitted on the quarterdeck and poop. Stowing the hammocks above the bulwarks provided some protection from small arms fire and splinters. Surprisingly, it was not until 1746 that hammock cranes were required to be fitted on all ships.

Right: The beakhead bulkhead is striking in black with varnished wood mouldings and trim. Small glazed lights are fitted into the square frames between the fluted columns decorating the face of each roundhouse. The positioning of the hawse holes is unusual since they exactly pierce the eking of the cathead bracket. It is remarkable how many strange and contrasting curves have been precisely worked into innumerable small wood pieces to produce so beautiful a form.

Broadside view showing the well balanced but relatively modest scale of the head and quarter galleries on this well-armed two-decker.

CHAPTER 9

A Queen Anne 3rd rate c1710

~ Acquisition ~

THOSE FEW MODELS, LIKE the present example, that retain their original display cases have a special ability to evoke the domestic settings for which they were originally intended. This one is equipped with launching poles that cannot be fitted while the model sits in its case. One can imagine the original owner unlocking the case and slipping the model out to set it up as though ready for launch. This lovely model first came to our attention in 1971 with the publication of Guy R Williams' book *The World of Model Ships and Boats*. Two photographs of the model are reproduced, and they were supplied by Sotheby & Company. We quickly learned that the model had been sold at auction in 1968, but we had no clue as to its whereabouts, and it earned a spot at the top of our search list.

At about that time, Henry moved to California to attend medical school at Stanford in Palo Alto. Henry was still there in 1975 when he discovered that there was a local ship model club, and when he joined he asked the president if he happened to know of anyone with a dockyard ship model. Henry was sceptical when he heard the response that, yes, one of the members did have a fine model. He was given the member's name and address, which was in neighbouring Menlo Park. In scarcely five minutes Henry was there, examining this very model. Unfortunately for us, the owner liked it as much as we did and was not ready to part with it. Patience proved a virtue, however, and in 1992 we received a call that the model was available, whereupon we bought it.

~ Provenance ~

The original owner was likely to have been James, 1st Earl Poulett, who was First Lord of the Treasury of Queen Anne for the years 1710–11. In 1968 it was sold from the collection of Earl Poulett

The original case for this model is a rare example and retains its original hand-blown glass panels. The front and back sides of the case are hinged below and drop down when unlocked to permit unobstructed study of the model.

The model is equipped with launching poles, even though they cannot be erected while the model is displayed inside its case.

A single seat of ease is positioned between each main head rail and the beakhead bulkhead. The chimney flue is visible on the foredeck. The anchor lining and billboard can be seen extending below the forechannel.

of Hinton St George, Somerset, along with other contents of Hinton House at a Sotheby's sale in London. It was purchased by the esteemed antique dealer, Ronald Lee, who sold it to an American collector living in California. We acquired the model in 1992.

~ Description ~

CONDITION

This model and its case have survived together, in original condition. The beautiful walnut display case is one of a very small number of tabletop cases to have survived from the early eighteenth century. Included with the model are a set of flagstaffs that would have been fitted in the mast holes at the time of launching, although the model needs to be removed from its case for these to be erected. The original dolphin carved cradle supports are slotted to fit between futtocks on the model, and they are attached to a slate baseboard with wooden pegs.

The restraint in decoration imposed by the Admiralty order of 1703 is clearly in evidence in this bow view. Carving has been reduced to only the figurehead and trailboard, and even the companion figure often found on the back of the lion has been replaced by a simple gilded panel.

CONSTRUCTION

Scale: 3/16in =1ft Hull length: 31in

This model is from the transitional years of the early eighteenth century when carved work on ships was being reduced and painted decoration was increasing. This is most evident in the plain head timbers and brackets and the simple gilded mouldings around the quarterdeck gun ports in lieu of carved wreaths. There remains, nevertheless, a fair amount of decorative carved work that is finely wrought and gilded, including lions, cherubs, dolphins, sea shells, and most notably a carved coat of arms centred on the taffrail. These are the arms Queen Anne used after the union with Ireland in 1710–14. Consistent with decorative conventions of the time, elegant scalloped shells adorn the upper finishing of the quarter galleries and surmount a pair of lights cut through the taffrail. Painted foliate decoration adorns the frieze planking and upper counter of the stern.

The gun deck is completely framed with two tiers of carlings on either side of the hatches and is supported by standard knees.

A QUEEN ANNE 3RD RATE c1710

Simple gilded circular mouldings surround the quarterdeck gun ports, in compliance with Admiralty Board instructions to reduce carved work on warships, and the topside friezes sport painted decoration for the same reason. The breast rail of the open gallery is supported by precisely carved balusters, curved to align horizontally when viewed from the waterline. A gilded scallop shell canopy forms the upper finishing.

The quarterdeck ladder provides access to the waist from both the gangway and the topside boarding steps, and the metal entry port stanchions are pierced for a hand rope to assist in climbing aboard. The ladder treads are concave in profile at the gangway, and gradually become convex towards the waist. Paired, turned balusters support the side gangway handrails, while single ones support the quarterdeck rail, and the removable rails of the central gangway are on metal stanchions. Barely visible on the model, this photograph reveals sheave slots on both the main jeer and top sail sheet bitts.

Abaft the mainmast, representational pumps have been fitted with pump dales rigged to the ship's sides. The fore jeer capstan is a drumhead type with six whelps supporting ten bars, and there is a small brass deck pawl on the port side. The main capstan also carries ten bars, which are fitted and project between finely turned pillars that help support the upper deck. There are pissdales inboard on both sides at the waist, and seats of ease are present on the beakhead, but at this time roundhouses have yet to appear on two-deckers.

This model has survived in an extraordinary state of preservation because it has remained in its original Virginia walnut display cabinet. The wood used in its construction is a type that originated in North America, but was grown in England as early as 1656. It was valued because of its rich colour and strength and was popular in the early eighteenth century as a material for fine furniture. This unique case retains its original hand-blown glass, and the front and rear panels are fitted with shell-form brass hinges and locks that allow them to fall open for closer inspection of the model. The superb carved and gilded dolphin cradles fit into a slate baseboard, and the whole case is supported on bun feet. For nearly 300 years, this model has been a focus of attention and interest in its owners' homes.

Literature

The following references include photographs and descriptions of this model:

Davis, F D, 'Talking About Sale-Rooms', *Country Life* (5 December 1968), pp. 1464–5.

Kriegstein, Arnold and Henry, 'The Kriegstein Collection of British Navy Board Ship Models', *Nautical Research Journal*, 38, 4 (December 1993), pp. 223–4, plates 13 and 14.

Williams, Guy R, *The World of Model Ships and Boats* (New York: G P Putnam's Sons, 1971), p. 14, 15, 295.

Wilson, P, ed, *Art at Auction 1968–69* (New York: The Viking Press Inc. 1969), p. 342.

Historical Perspective

It is possible to propose a tentative identification for this model. The size of the ship represented corresponds almost exactly with a number of small 3rd rates built during the reign of Queen Anne. The gun count is correct as well, although the standard distribution for this class had 24 guns on the main deck and 26 on the upper deck, which is reversed on this model. A deviation like this may well have occurred when some of these ships were built, however. Of the five examples constructed during Queen Anne's reign, two were launched in the 1710–14 period. One was the *Rippon* of 1712, but plans exist for this ship at the NMM and it does not match our model. The only other possibility is the *Lion* built by Rosewell and launched at Chatham in 1710. This may well be the man-of-war represented by our model, but no plans or depictions of this ship have yet come to light.

The painting of draped curtains and tassels along the lower counter is an early example of a convention that became quite popular in the second quarter of the eighteenth century.

LAUNCHING OF MEN-OF-WAR

Preserved with the model are a set of flagstaffs that would have been fitted at the time of launching. Wooden warships were built in waterside dry docks or slipways and launched by floating on a high tide or sliding into the water in a supporting cradle. Heavy furnishings such as ballast, ordnance and ammunition were brought on board later, and so too were the masts, spars and rigging. A mast could weigh over 12 tons, and launching a wooden warship was a tricky business even without the destabilising effect of these massive and lofty timbers, so they were added after the vessel was afloat. It is characteristic that in this age when even cannon were ornately

There is a delicately carved Queen Anne coat of Arms surmounted by a crown in the centre of the taffrail, and above each of the flanking stern lights there is a scallop shell. Completing the taffrail decoration, there are seated cherubs, the one to port carrying a twig of laurel and to starboard a sword. The quarter figures consist of crouching lions. There are hinged double doors in the centre of the screen bulkhead providing access to the state room.

decorated, no aesthetic concessions were made when launching warships. In lieu of counterbalancing masts and rigging, flagstaffs were fitted into each mast aperture. The ensign, jack and pennant were flown, and a small number of well-preserved Navy Board models also bear these launching flags. In unrigged models, these colourful additions enhance the overall decorative effect. Launchings were festive occasions, and Royalty were occasionally in attendance. An early example of this occurred on 24 September 1610 when King James I and a large retinue of courtiers turned out to witness the launch of the *Prince Royal* at Woolwich.[1]

The *Edgar*, approximately the same size as the subject of this chapter, was built by Randall and launched at Cuckhold Point, Rotherhithe, on 16 November 1758. This painting, by John Cleveley the Elder, memorialises the event. John Cleveley worked as a shipwright at the Deptford dockyard, where he may have been employed to paint the decorations on newly built men-of-war. He also painted detailed views of ship launches on the Thames based upon first-hand observations. Here the *Edgar* has just been floated and is still attached to her launching cradle; the spurs still bolted to the hull fore and aft. She is being towed to a mooring on the Thames and proudly flies her ensign, pennant and jacks. Poles for similar flags were originally fitted on four of our models (see chapters 2, 9, 10, and 14).

A QUEEN ANNE 3RD RATE c1710

The quarterdeck bulkhead is recessed aft of the break of the quarterdeck, and a central gangway extends to the mainmast. The baroque gilded dolphin cradles are attached to a slate baseboard by wooden pegs.

CHAPTER 10
The *Royal Oak*, 3rd rate of 1713

Acquisition

WE FIRST LEARNED OF this model when it appeared for sale at Sotheby's on 3 May 1995. It bore new and incorrect rigging, and although the hull sections had been screwed together, we could tell that it was built to separate above the gun deck. After purchasing the model, we had Philip Wride remove the rigging and separate the two parts of the hull. The painted work was then carefully conserved.

This model was known to be in the Royal Collection at Windsor Castle until 1902. Some years prior to buying the model, Arnold had come across a nineteenth-century engraving of the Queen's drawing room at Windsor Castle.[1] Published on 1 August 1816, it clearly showed a model of a 3rd-rate warship with launching flags in a glazed case on a side table. This is the earliest known depiction of a ship model in a private home, albeit a castle. After buying the *Royal Oak* model, we examined the print again, and the model depicted is clearly this one. It agrees in all visible details including size, double wales, colour and age. The late Queen Anne date of the model is independently confirmed by the design of the original case, since missing and probably sacrificed when the launching poles were exchanged for full rigging in the nineteenth century. The case is constructed in the same style as the original one for our model from Hinton House (see Chapter 9) which can be dated to 1710–14. This agrees exactly with the *Royal Oak*'s date of 1713. No such model remains at Windsor Castle, and the fortuitous existence of this pictorial evidence gives a unique view of the decorative use of this model in its early nineteenth-century royal home.

Left: This perspective view gives a good idea of the colour scheme of this model. Aesthetics were evidently of some importance to the shipwrights responsible for this ship. Even the gun port lids on the beakhead bulkhead have lions' faces painted on the under surfaces. The fiferail above the bulkhead has apertures to allow the foremost fore castle guns to be trained forward.

Provenance

Although the original owner of this model is unknown, it entered the British royal collection at Windsor Castle prior to 1816 and was likely a gift to King George. It remained there until it was transferred to the Royal United Services Museum, Whitehall, London, by command of His Majesty, King Edward VII in 1902. With the closure of that museum and the disbursement of its holdings, this model was sold in 1967 and purchased by the Glenbow Museum and Library in Calgary, Canada. It was never displayed there and was sold again in 1995.

This shows the model as it appeared when we purchased it. The original launching poles and flags had disappeared, and the model was ludicrously rigged. The two sections of the hull were also screwed together.

HISTORIC SHIP MODELS

This is the earliest known model that was built to separate at the gun deck. As a result, the hull has a 'solid' appearance in contrast to the open framing found on other models of this date.

This is a detail taken from an engraving published on 1 August 1816, showing the Queen's drawing room at Windsor Castle. *The Royal Oak* model can be seen in its original case, sporting launching flags. The model remained in the royal collections at Windsor until it was given to the Royal United Services Museum by King Edward VII in 1902.

~ Description ~

CONDITION

When acquired, there were some areas of the delicate japanning that had been worn and these were carefully restored, and other areas were conserved and protected beneath new 'varnish'. The nineteenth-century rigging was also removed. The model has otherwise descended through three centuries in extraordinarily good condition and is complete and original.

CONSTRUCTION

Scale: 1/48 Hull length: 44½in

This important early eighteenth-century Admiralty Board model is an example of the rare type designed to separate at the lower gun deck to reveal interior fittings and compartments and is the earliest example of this type known. It is likely that this model is the work

The framing of the lower deck forward includes the beams and carlings, along with the riding bitts, their shaped standards and associated scuttles. Bulkheads can be seen for cabins and storerooms on the orlop deck, whose beams and carlings are also visible.

of Hayward, a master shipwright working at the Woolwich dockyard in the first quarter of the eighteenth century. In addition to this model of the *Royal Oak* of 1713, three other models of ships built at Woolwich (*viz* the *Revenge* of 1718, now in the Thomson collection at the Art Institute of Ontario, but formerly in the Kriegstein collection; the *Britannia* of 1719, now at the NMM; and the *Royal George* of 1715, now at the Hohenzollern Museum, Germany) are all of this rare 'sectional' type and appear to be the

The detailed construction and operation of the lower deck pumps can be carefully studied when the upper section of the model is removed.

The main bitts include sheaves for the braces and topsail sheets. The gun carriages are complete with trucks and bolsters, in contrast to the earlier model cannon that were fitted with simple sleds.

work of the same master craftsman. Significantly, all are built out of white pine, an unusual feature that further relates them.

Beneath the waterline, the hull is constructed of pine timbers that have been carefully formed and hollowed out to a thickness of ¼in. Draught of water marks are inscribed on the stem and sternpost and indicate that she drew 16ft when fitted out. The orlop and gun decks are fully framed with two rows of carlings let into the beams with ledges let into angled scores on the carlings. The outer tier of carlings are covered with plank decking on the orlop, except for the wings where two planks are laid across the beams to indicate the carpenter's walk. This runs fore and aft 'twixt wind and water', and formed a passageway that gave access to the hull at the waterline in time of battle to allow quick repair of shot holes. Various storerooms, passages and compartments are shown on the orlop deck, and a finely built stair provides access to the cockpit. The gun deck beams are supported by lodging and hanging knees, and there are breast hooks fitted at the bow and transoms at the stern. The usual ridding bitts are fitted along with a drumhead main capstan. The pump well has solid walls below the orlop level and is pierced with ventilating holes above, conforming to normal practice.

The pumps are constructed in great detail and are unusually sophisticated. There are paired chain pumps, operated by metal winches that seamen would crank to operate the chain. This consists of a long loop of linked metal segments and attached at fixed intervals there is a series of leather valves or dishes. The diameter of these valves fits snugly into the bore of a pair of wooden tubes that run from the well in the hold up to a wooden cistern on the gun deck. As the metal rollers turned, sprocket wheels on the shaft of the roller and in the well would operate the chain to lift water into the cistern. From the cistern, gravity would carry the water through wooden tubes called 'dales' to holes in the side of the ship, midway up the port cill. The working of this apparatus can be studied in detail in this model. In addition to the chain pumps, a pair of suction pumps are also fitted. These operate through their own wooden tubes, which empty into the same cistern abaft of the chains via short inclined wooden pipes. The pumps are located abreast of the mainmast.

The model is armed with a complete set of well-detailed wooden guns on carriages with trucks. Because it separates at the level of the gun deck, this posed a problem for the model maker as

The head grating on the port side consists of ledges only, while on the starboard side fore and aft battens have been let in to complete the grating. Inside each roundhouse there is a seat of ease and a doorway through the bulkhead for access.

the gun trucks rest on the deck, but the barrels extend out through the ports, whose cills are attached to the upper section. This dilemma was solved by attaching a pair of metal pins to the front of each gun carriage, which insert into holes in the lower cills. Thus, the guns are not attached to the deck and lift up with the topsides.

The upper deck is full of interesting details including a galley stove beneath the forecastle, pissdales in the forward part of the waist, and bulkheads creating numerous cabins and passageways with doors and lights. Carlings are let into the deck beams and are rabbeted for an inner tier of ledges. The outer tier is replaced by a solid plank that supports the upper deck guns. Hatches on the deck are pierced with square holes, and the coamings are painted black. Numerous sheaves are set into the topsides along with glazed lights.

The head is an area of special interest. Inboard of the beak bulkhead are a pair of footstools leading to the doorways, and there is separate access to the roundhouses, which are fitted with seats of ease. The beakhead gratings are unusual. There are the usual gratings between the headrails at the level of the platform, but the port side is fitted only with athwartship ledges, whereas the starboard side is fully built with fore and aft battens let into the ledges. The carlings that support these gratings are let into a fore and aft beam reinforced with lodging knees, and this forms the forward termination of the grating. There is another grating, however, which extends out to the dead block for the fore tack, centred on the third head timber but at the level of the upper gun deck. This has also been built up in asymmetric fashion, this time with fore and aft battens on the port side and athwartship ledges to starboard.

The quarterdeck has a large wheel immediately aft of the mizzen, and there are trumpeters' cabins on the poop deck. The securing bracket for the ensign staff is attached to a decorative

This aerial view shows the arrangement of the forecastle deck, quarterdeck, poop, flag lockers and upper deck at the waist.

View of the quarterdeck aft showing the wheel and behind it the doors to the master's cabin and the captain's sleeping quarters. The guns are made of wood, and the turned barrels and tompions are painted gold while the muzzles are red.

The taffrail decoration of this model confirms its identity, as it features a bust of King Charles II (partly obscured by the crane for the central lantern) surrounded by oak leaves and branches. On either side are cherubs blowing trumpets, a crowned medallion with a CR cipher on the port side and a crowned anchor in a medallion to starboard. There are two open galleries, and beneath each breast rail are solid panels pierced for four gun ports with camouflaged lids hinged to open downwards.

Roundhouses and/or seats of ease appear on almost all naval ship models, but urinals or pissdales are much less common. All four of our Queen Anne period models have them, and they are all fitted on the upper gun deck against the inboard side of the bulwarks in the forward portion of the waist. Each model features a different design, and those fitted on the *Royal Oak*, illustrated here, would have been made of lead or copper. Pissdales were a great convenience to the crew, who would otherwise have had to make their way to the head to relieve themselves. Contemporary ship models offer the best evidence for the placement and appearance of these practical fixtures.

support, and the lantern cranes are braced by wire stays. The most arresting aspect of this model is the ornate painted and carved decoration, which is in chinoiserie. The carving includes oak leaves on the torso of the lion at the head, an elaborate taffrail incorporating dolphins, angels with trumpets, crowned medallions with the royal cipher and Admiralty Anchor, and figures beneath a central oak tree. The painted decoration reinforces the same theme with a continuous frieze stem to stern of entwined oak leaves. This continues around the stern on the upper cove. The upper frieze is covered with lions' heads, trophies of arms, guns, cannon and swords, and mandarin figures. There are two open stern galleries, and each breast rail is pierced for four guns with painted lids hinged below.

The underside of each gun port lid sports a painted oriental-style lion's face, at least some of which have a ferocious look.

Literature

The following references includes photographs and descriptions of this model:

Marshall, Percival, 'Ship Models at the Royal United Services Museum', *Ships and Ship Models* London, Percival Marshall & Co, Vol. 1, No. 1, September 1931, p. 12.

Walker, Grant H, *The Rogers Collection of Dockyard Models*, Vol. 2 (Florence, OR: SeaWatch Books, LLC, 2018), p. 129.

Exhibitions

London, The Royal United Service Museum, 1902–67.

Greenwich, National Maritime Museum, *Ship Models from the Great Age of Sail 1600–1850*, 18–20 April 1996.

There are fifty-eight leonine faces painted on the inside of the gun port lids, each with a different expression.

Historical Perspective

OAK LEAVES CONCEAL A KING

The first *Royal Oak* was built as a 76-gun 2nd rate at Portsmouth dockyard and launched in 1664. She was burned at Chatham by the Dutch during their daring raid in June 1667. A new *Royal Oak* was built as a 70-gun 3rd rate and launched at Deptford in 1674. She was rebuilt at Chatham in 1690, and again in the form of this model, at Woolwich in 1713. This is therefore the third *Royal Oak* and was built by Acworth.

Complex religious and political considerations continued to embroil England in a series of conflicts with her European neighbours throughout much of the eighteenth century. Shifting alliances pitted the Royal Navy against a variety of other naval powers. On 11 August 1718, the *Royal Oak* saw action against the Spanish at the Battle of Cape Passaro and later helped thwart the Spanish siege of Gibraltar in 1727. On 11 March that year, she captured the newly built *Nuestra Senora del Rosario* before a four-part treaty brought a temporary end to hostilities. She was broken up in 1737.

The name *Royal Oak* was not a reference to the material favoured by shipwrights for building their war machines, nor did it refer to the loyal heart of the British tar. Rather, it commemorated a particular tree in the woods near Boscobel that helped preserve Charles Stuart for his restoration to the throne. A descendant of this tree can still be seen near Boscobel House, and its progenitor played a role in the most remarkable escapade in European political history.

In 1646, with England in the throes of civil war, embattled King Charles I sent his eldest son, Charles Prince of Wales, to France for safety. After Cromwell's Parliament executed the King, Charles' son led the first attempt at restoring the monarchy, which became the second Civil War. Charles gained favour in Scotland by agreeing to abandon Anglicanism, keep the Church of Scotland Presbyterian, and spread Presbyterian reforms to Ireland and England as well. On 5 February 1649, Charles II was proclaimed King of Scots in Edinburgh, and he returned from France to land in Scotland on 23 June 1650. He was officially crowned King of Scots at Scone, Perthshire, on 1 January 1651 and raised an army to confront Cromwell. Enjoying initial success, his forces pushed south into England as far as Worcester, where the second Civil War came to an end on 3 September 1651. Charles II and his Scottish forces fought bravely but were defeated by the Parliamentary army

This head, painted on the quarterdeck frieze planking, has an oriental look, consistent with much of the rest of the decoration of this model. The oak leaves on the plank below run all around the model. The sheaves fitted into the plank above the main channel are not often shown and are an indication of the detailed work on this model.

led by Cromwell. The Battle of Worcester had been hard fought and Charles was lucky to be alive at the end of the day, but his continued survival was highly unlikely. The Commonwealth Period had begun, and Charles II was a fugitive in his own land. For the next six weeks Charles Stewart, 'two yards tall' and with a price upon his head, managed to elude his pursuers in a remarkable cat and mouse chase through south-west England that remains 'the most stirring and romantic story in the chronicles of the English throne'.

Many versions of the escape have been written, but the most authoritative is one recorded by Samuel Pepys in shorthand from the King's own narration. Pepys first heard the tale from the King himself on board the *Royal Charles* on 23 May 1660, while accompanying him back to England from Holland. In his diary entry for that day, Pepys included highlights from the tale, but he did not inscribe a complete record until years later, when on the night of Sunday, 3 October, and continuing on Tuesday, 5 October, the King narrated the complete story for Pepys to record in shorthand. The first printed version of this recitation did not appear until 1825, and is the basis for the following synopsis.[2]

On the night of 3 September, the battle being hopelessly lost, the King escaped from Worcester in the company of several loyal noblemen and a contingent of cavalrymen. 'But we had such a number of beaten men with us (of Horse) that I strove as soone as ever it was dark to gett from them. And though I could not gett them to stand by me against the enemy, I could not gett ridd of them now I had a minde to it.' The King and his reduced entourage travelled all night and arrived at a Catholic safe house called Whiteladies (a former priory) at dawn. The King was urged to attempt to reach Scotland, which I thought was absolutely impossible, knowing very well that the Country would all rise upon us, and that men who had deserted me when they were in good Order would never Stand to me when they have been beaten.

    This made me take the Resolucion of putting my selfe into a disguise, and endeavouring to gett a Foote to London in a Country-Fellowes habbit, with a pair of ordinary grey Cloath Britches, a Leathern Dublett and a greene Jerkin which I tooke in the House of White-Ladyes. I also cutt my Haire very short, and flung my Cloathes into a Privy-House, that noe Boddy might see that any body had beene Stripping themselves, I acquainting none with my Resolucion of goeing to London but my Lord Wilmott, they all desireing me not to acquaint them with what I intended to doe, because they knew not what they might be forced to confess: On which Consideracion they with one Voyce begged of me not to tell them what I intended to doe.

This precaution proved wise, as after the King parted with his entourage, most of the noblemen were captured including Lord Buckingham, who was condemned to death, and Lauderdale, who was imprisoned. But Charles left Whiteladies with a 'Country Fellow' named Richard Penderell and hid in a nearby wood without food or drink. There they saw a troop of soldiers looking for fugitives, but it rained all day, which kept them from searching the woods. The King questioned Penderell about sympathisers who could help the King reach London, but he did not know any, which prompted Charles to change his plan and try to reach the Welsh coast and head for France.

    Under cover of darkness, the two set off on foot, but were accosted by a miller, who challenged them and bade them, 'Stand, or else I will knock you down,' but they ran off with some men in pursuit. When they could not run further, they leaped over a hedge and lay still for half an hour, losing their pursuers. They made their way to the house of a Mr Woolfe, who agreed to hide them the next day, but the safest spot was in the barn 'behinde his Corne and Hay'. Discoursing with Mr Woolf and his son that night, the King learned that soldiers were guarding the River Severn and that crossing into Wales was impossible; the plan to head for London was revived. At dark, they set out to hide at the house of one of Penderell's brothers, but were advised that the house would be too dangerous, and Penderell made the alternative suggestion,

> that he knew but one way how to pass the next day, and that was to get up into a greate Oak in a pretty plaine place, where we might see round about us; for the Enemy would certainly Search all the Wood for People that had made their Escape. Of which Proposicion of his I approving, we went and carried up with us some Victualls for the whole day, vizt, Bread, cheese, Small Beere, and nothing Elce, and got up into a greate Oake that had been Lop't some 3 or 4 Yeares before, and being growne out again very Bushy and Thick, could nott be seene through. And here we stay'd all the day …

While Charles and Richard Penderell were in the oak, they saw 'soldiers goeing up and downe in the thickest of the Wood, searching for persons escaped, we seeing them now and then peeping out of the Woods'.

    That night Charles climbed out of the tree and continued his harrowing journey that would last another six weeks. With a £1,000 price on his head, and despite his imposing height, distinctive black hair and royal demeanour, he managed to brazenly travel about the countryside and defy his pursuers. Charles changed his disguise to that of a serving man, attendant upon Jane Lane, who was off to visit her pregnant sister. The King quickly adapted to the clothing, diet and discomfiture of his erstwhile common subjects, suffering hunger, blistered feet and subservience with remarkably high spirits.

    Pepys describes several moments of comic melodrama. While on the road to Longmarson (now Long Marston), the King's mount lost a shoe, and while Charles was holding his horse's foot for the smith he asked,

> What Newes? He told me that there was noe newes that he knew of since the good newes of the beating of the Rogues, the Scotts. I asked hem whether there was none of the English taken that joined with the Scotts. He answered that he did not here that that Rogue Charles Steward was taken, but some of the others he said were taken, but not Charles Steward. I told him that if that Rogue were taken he deserved

to be hanged more then all the rest for bringing in the Scotts. Upon which he said that I spoake like an honest man, and soe we parted.

On 13 September Charles ate breakfast in the buttery at Abbot's Leigh,

and as I was sitting there, there was one that looked like a Country fellow satt just by me, who talking gave soe particular an Account of the Battle of

The open stern galleries extend to the quarter where the breast rails become the gallery rims. The quarter pieces consist of a figure blowing a horn above, with a cupidor riding on a dolphin below. A series of four scuttles are placed between the guns beneath the quarterdeck to admit extra light to the cabins. The glazing is mica and the mullions are gilded.

Worcester to the rest of the Company, that I concluded he must be one of Cromwells Soldiers. But I asking him how he came to give soe good an Account of that Battle, he told me that he was in the

King's Regiment. By which I thought he meant one Coll. Kings Regiment. But questioning him further, I perceived that he had beene in my Regiment of Guards in Major Broughtons Company, that was my Major in the Battell. I asked him what a kinde of man I was, to which he answered by describing exactly both my Cloathes and my Horse and then looking upon me he told me that the King was at least 3 fingers taller than I. Upon which I made what hast I could out of the Buttery, for feare he should indeed know me, as being more afraid when I knew he was one of our owne Soldiers then when I took him for one of the enemys.

Later on, Lord Wilmott was sent to Lyme to try and arrange transportation across the Channel, and Charles was to meet up with him in Burport (now Bridport).

Soe Franck Windham and Mrs Conesby and I went in the morning on Horse-back away to Burport, and just as we came into the Towne, I could see the Streets full of Redd-Coates, Cromwells Soldiers (being a Regiment of Coll. Haynes's 1500 men going to imbarke to take Jerzey) at which Franck Windham was very much startled, and asked me what I would doe. I told him that we must goe impudently into the best Inn in the Towne and take a Chamber there, as the only thing to be done; because we should otherwise miss my Lord Willmott in case we went any whether elce, and that would be very inconvenient both to him and me. Soe we Rodd directly into the best Inn of the place and found the Yard very full of Soldiers. I alighted, and takeing the Horses thought it the best way to goe blundering in among them, and lead them through the middle of the Soldiers into the Stable, Which I did and they were very angry with me for my rudeness.

As soon as I came into the Stable I tooke the Bridles off the Horses, and called the Ostler to me to help me give the Horses some Oates. And as the Ostler was helping me to feed the Horses, Sure, Sir (Sayse the Ostler) I know your face. Which was noe very pleasant Questian to me, but I thought the best way was to ask him where he had lived? Whether he had alwayes lived there or noe? He told me, that he was but newly come thether, that he was borne in Exeter, and had been Ostler in an Inn there, hard by one Mr Potter's, a Merchant, in whose House I had laine in the time of War. Soe I thought it best to give the fellow noe further occacion of thinking where he had seen me, for feare he should guess right at Last. Therefore I told him, Friend, Certainly you have seene me there at Mr Potters, for I served him a good while, above a yeare. Oh, says he, then I remember you a Boy there, and with that was putt off from thinking any more on it but desired that we might drink a Pott of Beere together. Which I excused by saying that I must goe waite upon my Maister, and gett his dinner ready for him, but told him, that my Maister was goeing for London and would returne about three Weekes hence, when he would lye there, and I would not faile to drink a pott with him.

Through numerous close calls and despite constant danger, the King displayed unfailing courage and finally left England at 4am on Wednesday, 15 October, departing from Shoreham in a small brig named *The Surprise* and landing near Rouen in France the same day.

After Charles' triumphant return to England nine years later, the Royal Oak came to symbolise the indomitable spirit and resourcefulness of both the people and their leader. The Royal Navy launched its first *Royal Oak* at Portsmouth in 1664, and her name has been glorified through generation after generation of warship christened in memory of those seminal times. The latest *Royal Oak*, the eleventh to bear the name, was a battleship built in 1916 and sunk by a German submarine on 14 October 1939.

References

Fraser, Antonia, *Charles II* (New York: Alfred A Knopf Inc, 1979).

THE *ROYAL OAK*, 3RD RATE OF 1713

When the upper half of the model is lifted, the main deck guns remain in position by pins that hold the carriages to the bulwarks. All external surfaces of the *Royal Oak* are painted, and the decoration is in chinoiserie style. A number of Admiralty models from the first half of the eighteenth century feature oriental designs, and Chinese craftsmen may have had a hand in producing them. It is possible that the actual ships also exhibited this artistic fashion.

The silk lines that elevate the sections of these two models can be seen attached to cleats at the edge of the backboard. The stem/figurehead elements are perpendicular to the frame sections and help lock them in place when erected. The *Greyhound* model is in the foreground.

CHAPTER 11

The *Diamond*, 5th rate of 1723, and the *Greyhound*, 6th rate of 1720

Acquisition

IN OVER THREE DECADES of searching for seventeenth- and eighteenth-century British dockyard ship models, we encountered only three paper or card ship models made at the royal dockyards. These were fascinating and delicate pop-up models, each folded into a book binding accompanied by a draft of the same ship. Paper models are even more fragile than the more common wooden versions. In fact, one seventeenth-century example was so ephemeral that it had entirely disappeared; only its paper keel supports remained.

We were therefore surprised to learn of a folio-size binding containing two pop-up models that was featured on an episode of the British TV programme *Antiques Roadshow*. This folio was in the Royal United Services Institution library at the time, and we eventually bought it in a private sale in 2009.

Provenance

The drafts that accompany these models bear the Cypher and Coronet of Prince William Augustus, the youngest son of King George II, and are dated 15 April 1731. This is just two days after the Prince's tenth birthday. According to an old label affixed to the binding, the folio remained in the royal collections until it was given to the Royal United Services Institution Museum by King George V. The label reads: 'This Book was presented to the Royal United Service Institution by the Librarian, Royal Library Windsor Castle. Deposited by H.M. The King 10 July 1922.' Upon the final dissolution of the museum in 1965, it was transferred to the RUSI library and was de-accessioned in 2009.

Description

This is a portrait of Prince William Augustus, Duke of Cumberland, with his mother Queen Caroline. William is ten years old in this painting, the same age as when he would have received the pop-up models.

A brown leather and gilt-tooled presentation folio opens to plans and folding skeletal models of two naval vessels, the *Greyhound* and the *Diamond*, dated 1719 and 1723, bearing the coronet and cipher of Prince William Augustus, son of King George II, dated 15 April 1731.

The day 13 April 1731 was the tenth birthday of Prince William, whose titles included Duke of Cumberland, Marquess of Berkhamstead, Earl of Kennington, Viscount of Trematon, and Baron of the Isle of Alderney.

CONDITION

In view of the fragile nature of the pop-up models, which were built entirely of moving parts and intended to be handled, it is surprising that they have survived nearly 300 years relatively intact. When we acquired them, there was no evidence of significant restoration, and only the stem section of the *Greyhound* and the beak bulkhead section of the *Diamond* were missing. The models were cleaned by Clare Reynolds, and the missing sections were replaced and minor paint touch-ups undertaken by Philip Wride.

CONSTRUCTION

Dimensions: 29in × 21in closed

The hand-drawn lines, profile, half-breadth, and body plan of two naval vessels, the *Diamond*, a 5th rate, and the *Greyhound*, a 6th rate, are on the inside front cover of this leather-bound wooden folio. Mounted on the inside of the back cover are pop-up models of each vessel built to the same scale and constructed of cross-sectional cutouts that are erected by pulling on specially rigged strings. The leather folio is hand-tooled with gilt embellishments that are echoed in the border of the drafts mounted on the inside cover. The drafts are finely drawn in ink and watercolour, and show the exterior appearance and decoration of both vessels, including oar ports on the *Greyhound*. Launching flags are displayed on all views, and the stations, which are represented by cross-sections on the models, are lettered and numbered in the standard manner on the drafts. Scale markings and principal dimensions of both ships appear in cartouches, and the cipher of Prince William appears surmounted by a coronet and supported by tritons. Beneath his cipher is the date 15 April 1731.

The models corresponding to the drafts are mounted on the inside of the back cover. Each is composed of twenty-four cross-sections or frames, delicately cut out of laminated parchment and attached to the folio with special hinged supports. The sections are carefully painted on both sides and include guns, bulkheads, capstans, launching flags, and painted bow and stern decorations. The figurehead and cutwater are hinged at right angles to the other sections and help define the ship's profile when erected.

In each model the sections are linked by a pair of silk lines that pass through holes in the upper deck and keel. There is a small wooden toggle tied behind each frame so that when the lines are

The tooled leather binding gives no hint of the pop-up models within.

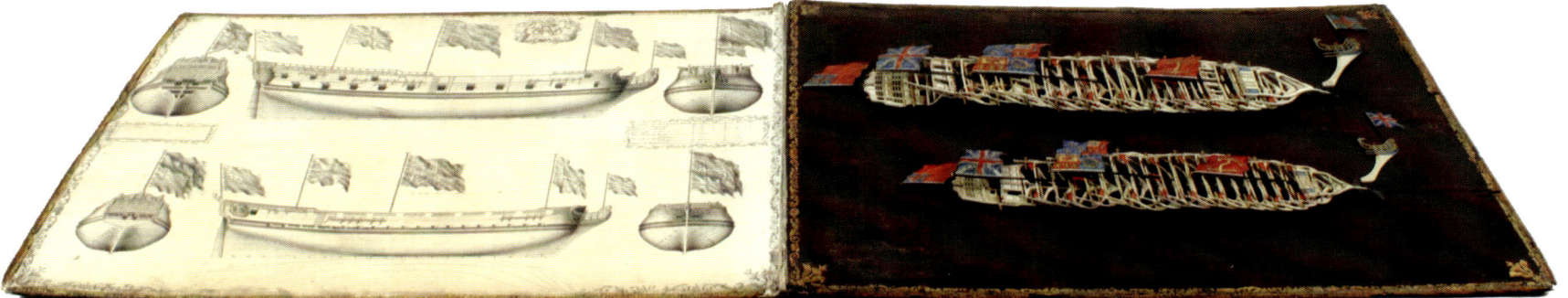

The opened folio reveals the drafts and models, which are all to the same scale and combined with the models convey an accurate impression of the three-dimensional appearance of the ships. The models are flat but can be erected with a tug on the silk lines running through the cut-out sections.

pulled the models pop up and a colourful, three-dimensional ship materialises. Wooden cleats are attached to the end board to make the lines fast.

A quarter view from the stern highlights the festive contribution that the launching flags and cannon make to the overall colour scheme. The 5th-rate *Diamond* is on the left.

Literature

The following reference includes photographs and a description of this model:

Pardy, Kary, 'The Maritime World Through Miniatures and Beyond', *Journal of Antiques & Collectibles*, Vol. XX No. 8, Sturbridge, Ma., Weathervane Enterprises Inc, November 2019, p. 39.

Exhibitions

Salem, Massachusetts, Peabody Essex Museum, 2019–present.

Historical Perspective

The *Diamond* represented here is the 1723 rebuild of the ship originally launched in 1708. She was rebuilt at the Deptford dockyard when Stacey was surveyor and launched on 13 March 1723. *Diamond* was a 40-gun 5th rate and carried a crew of 250.

The *Diamond* was in the West Indies for the blockade of Porto Bello, and later served under Vice Admiral Vernon on the Spanish Main. According to John Fanning Watson's collection of anecdotes and memoirs,[3] the *Diamond* engaged two pirate vessels while cruising in Honduras in 1725. The pirates Skipton and Joseph Cooper commanded the pirate vessels. One pirate ship was captured, but the captain and crew of the other blew themselves up rather than be taken.

While under the command of Captain Charles Knowles, she participated in the bombardment and capture of Fort San Lorenzo on 24 March 1740, netting over £70,000 in booty plus eleven brass guns. During two years in the West Indies, the fleet lost two flag officers, seven captains, fifty lieutenants and 4,000 subordinate officers and men from disease. In 1744 the ship was sold out of the Navy for £301.

The *Greyhound* was the twelfth vessel in the Royal Navy to bear that name and was also built under Stacey at the Deptford dockyard. Launched on 13 February 1720, she was a 6th rate carrying 20 6pdr guns on a single deck, with a crew of 140. Her early career paralleled the *Diamond* as both served in the West Indies and participated in the blockade of Porto Bello. While in service in the West Indies under captain Peter Solgard, the *Greyhound* was dispatched to find the notorious pirate Edward 'Ned' Low.[1, 2] On 10 June 1723, the *Greyhound* caught up with him. Low was sailing in the *Fancy* schooner, while his associate Charles Harris captained the sloop *Ranger*. The *Greyhound* pretended to flee as a ruse to draw the pirates closer and surprised them when she came about and delivered a withering broadside. The *Ranger* attempted to flee but was pursued and captured, while Low escaped in the *Fancy* and lived to continue his piratical exploits for several more years. The *Greyhound* sailed with her prize to Rhode Island, where Harris and his twenty-five-man crew were tried and condemned. The crew were hanged near Newport, Rhode Island, but Harris was dispatched to England and met his end at Execution Dock, Wapping. The exploit earned Solgard the Freedom Medal of the City of New York encased in a gold box decorated with scenes of the *Greyhound* engaging the *Fancy*. The *Greyhound* lasted twenty-one years and was broken up in 1741.

WHAT TO GIVE A PRINCE ON HIS TENTH BIRTHDAY

It is a mystery why models of the *Greyhound* and *Diamond* were chosen for presentation to the young Prince. However, the ships had much in common, as both were built at the same dockyard by the same master shipwright and both served in the West Indies and participated in the attack on Porto Bello. In addition, they both independently struck a blow against piracy during the heyday of the Caribbean buccaneers. Perhaps the association with pirates was meant to help capture the imagination of a ten-year-old boy.

Note the trompe-l'oeil gun port lids and stern lanterns and the way the gun deck can be viewed through the open chase ports. The knotted thread with toggle stop centred just above the counter is also visible.

THE *DIAMOND*, 5TH RATE OF 1723, AND THE *GREYHOUND*, 6TH RATE OF 1720

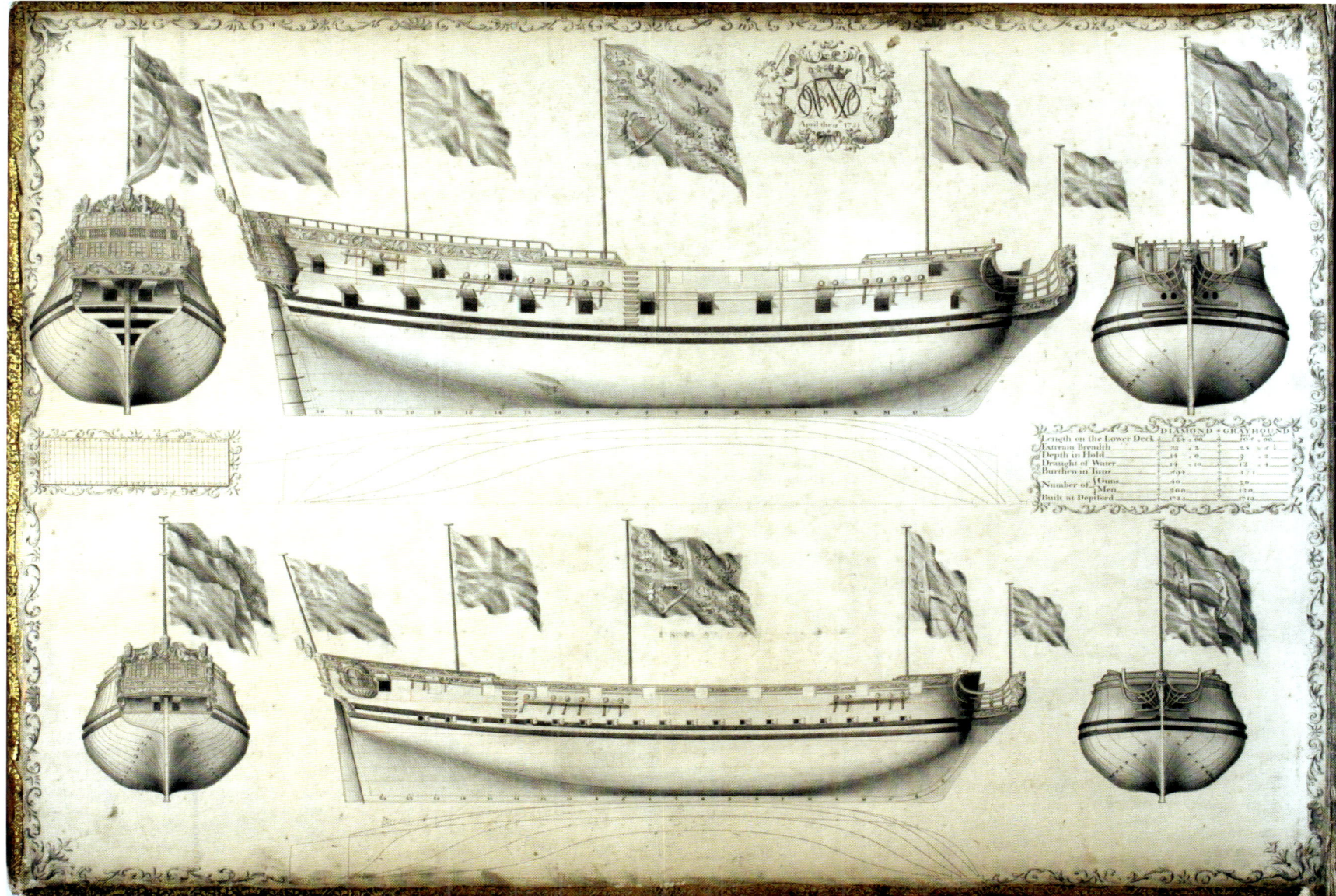

The carefully drawn lines profile, half-breadth and body plan of the *Diamond* (above) and *Greyhound* (below) include a scale and table of dimensions. The cartouche at top includes the coronet and cipher of Prince William Augustus, son of King George II, and the date 15 April 1731.

FOLIO MODELS: A RARE DOCKYARD ARTFORM

There is very little known about the practice of making cut-out or pop-up models of Royal Navy ships. To our knowledge there were two makers, the first of whom was William Keltridge. A ships' carpenter, Keltridge produced at least two volumes of drafts of naval ships around 1685. One of these, now in a library in Germany, originally featured a pop-up model mounted to the inside back cover of the folio, much as in our example. The model in this instance has sadly disappeared and only the telltale mounts for the sections remain in place. Another manuscript by William Keltridge, this one filled with invaluable details of ship construction, rigging, maintenance, etc, has fared better at the hands of time. When loaned to the National Maritime Museum some years ago, it had a pop-up model of a 6th rate mounted to the end papers opposite a draft of the same vessel, exactly in the style of our models, and also sporting launching flags. This late seventeenth-century example is the earliest instance of this novel artform known to survive.

A second practitioner revived this style of representation in the early eighteenth century and produced the models described in this chapter. He remains anonymous, but there is another example of his work at the National Maritime Museum, Greenwich, England. This is a pop-up model of the 6th rate *Garland* built at the Sheerness dockyard under surveyor Ward and launched in 1724. Like ours, it is in a self-contained folio, mounted alongside a draft of the same ship, and appears to be by the same hand as the *Diamond* and *Greyhound*.

This ends our very limited knowledge of the creators of these remarkable objects. Other examples must have been made, and we hope that additional ones will come to light over time and expand our understanding of this intriguing, esoteric practice.

THE BUTCHER OF CULLODEN

One of the most interesting features of this book of models is the monogram it bears of Prince William with the date of 15 April 1731. That this was two days after the Prince's tenth birthday and that the model descended in the royal family is, of course, no coincidence. William was the younger son of King George II and Queen Caroline of Ansbach, and from an early age HRH Prince William was groomed by his parents to become Lord High Admiral.

It was William's older brother, Frederick, whose son would become King, but it was William who received most of his parents' love and attention growing up. By age five, when he became the Duke of Cumberland, he was already Marquess of Berkhamsted, Earl of Kennington, Viscount Trematon, and Baron of the Isle of Alderney and was entitled to wear both the Red Ribbon of the Bath and Blue Ribbon of the Garter. His parents King George II and Queen Caroline ensured that William would have the best of tutors. These included Sir Jacob Acworth, Surveyor to the Navy, who may have had a hand in the production of this folio for the young Prince. In any event, it appears likely that the models entered the royal collections on the occasion of the Prince's birthday, and that the gift was intended to strengthen the boy's interest in naval matters. It may have worked, for in August 1740, at the age of nineteen and with England at war with France and Spain, William set off to join the Home Fleet.

Prince William, the Duke of Cumberland, was received on board HMS *Victory* under command of Admiral Sir John Norris. Things went badly from the start. A gale blowing from the west kept the squadron at anchor in Portsmouth for nine days. When the weather finally broke, the squadron weighed anchor, reaching Portland by nightfall. It was then that the look-out spotted an unknown vessel bearing down on *Victory*. The 1st-rate warship could not bear off in time, and a collision was unavoidable. The stranger struck *Victory* in the bow, carrying away her cutwater, and Cumberland was awakened by 'a universal scream and outcry close under his cabin window from the ship who had done the damage and who thought themselves sinking'.[4] By the light of day it was evident that the *Lion* had been the culprit (see Chapter 12). Twenty-eight men were lost aboard the *Lion*, and the squadron put about for St Helens. There the admiral moved his flag to HMS *Boyne* and accommodations were also found for Cumberland and his retinue in the smaller vessel. Upon setting off again, the squadron encountered another gale and this time they sheltered in Torbay. Meanwhile, plans for the squadron kept changing, and Cumberland grew impatient while messages were exchanged between the admiral and the Admiralty. Finally, in mid-September the Prince disembarked and returned to London, concluding a tedious and inauspicious naval adventure that proved to be his last. When a combined operation to the West Indies was finally organised, the King refused to allow his son to join for fears of his health. Just as well, for scores of men died from yellow fever, the expedition was a fiasco and nothing was achieved.

The Duke of Cumberland resolved to devote his future to the Army instead, and over the next several years he laid the groundwork for a distinguished military career. As a scion of the royal family, his meteoric rise through the ranks was no surprise, and by the age of twenty-one he was a major general. Cumberland, however, was on a collision course of his own with the forces of Charles Edward Stuart, the Young Pretender, which was to have historic consequences for all involved.

'Bonnie Prince Charlie,' the Young Pretender, laid claim to the throne of England for his father, James (the Old Pretender), in the name of his grandfather, King James II. Encouraged by expatriate Jacobites, Charles Edward Stuart left his exile in France and landed in Scotland on 23 July 1745, virtually alone (see Chapter 4). Over the ensuing months, he garnered support for his cause and fought across Scotland in a campaign that came to an end on Culloden Moor on 16 April 1746. By then he was supported by a force of some 5,000 Jacobites and faced an army of 8,000 government troops led by William Augustus, the Duke of Cumberland.

On the eve of the battle, Cumberland celebrated his twenty-fifth birthday, only fifteen years since he received our folio of models. The next day his name was to be forever inscribed in the annals of British history, to live on in a mix of fame and infamy. The Battle of Culloden was a decisive rout. Over in barely half an hour, 2,000 rebel soldiers lay dead or dying while only 350 government troops were lost. Bonnie Prince Charlie escaped but had no will to fight again. Cumberland, however, was not done yet. His objective was to abolish the clan system and ensure London's rule over the Highlanders. Wounded Jacobites were killed on the battlefield, while others were executed after being taken prisoner. Government dragoons were dispatched to hunt down surviving rebels, and they proved indiscriminate in their brutality. Reprisals against the rebels and their supporters continued until July, when Cumberland left Scotland to return to London. The role that HRH the Duke of Cumberland played in trying to control his troops and avoid excessive force is a matter of debate, but his name has ever since been tarnished with the sobriquet, 'the butcher of Culloden'.

Cumberland returned to London as a hero and acquitted himself well in the Army despite malicious and persistent allegations of barbarity during his campaign in Scotland. On 20 March 1751, his brother Frederick, the Prince of Wales died aged forty-four. He was survived by his wife Augusta and eight children. Frederick's eldest son, George, was only twelve years old, and the King wanted Cumberland to become Regent. William's reputation, however, mitigated against this and Parliament settled upon Frederick's widow to become the Princess Dowager Regent. Cumberland retired to Windsor, although he was politically active in the government of his nephew, who became King George III. William died unexpectedly on 31 October 1765, at the height of his political power and influence and, like his brother, aged forty-four.

The folio with both models in their upright positions. The 6th-rate *Greyhound* is on the left.

While this model was at the Royal United Services Museum, London, it was lent to the Science Museum in South Kensington to be photographed. Black and white photographs printed from the resulting glass plates were on sale at the museum for many years and were the first evidence we had for the existence of this model. It was not until after we purchased the model from a collector in Germany that we discovered that its lovely shape was complemented by an equally beautiful colour scheme.

CHAPTER 12

The *Lion*, 4th rate c1738

— Acquisition —

AMONG THE MANY SPLENDID models made in the eighteenth century, few can compare in terms of sheer beauty with the stunning examples of those decorated with lacquer work in the style known as chinoiserie. The delicate, fragile painted surfaces in rich tones of yellow, red, gold and black, belie the function of these ships as powerful war machines, which indeed they were. The Science Museum in London used to be a wonderful resource for model enthusiasts. In the 1970s when we first visited, there was an extensive library of photographs for sale, which were of ship models photographed at the museum but not necessarily in their permanent collections. Back then, many models were known only from these photographs, and one set was of particular interest. They depicted a beautiful and well-preserved early eighteenth-century two-decker bearing the name *Lion* in the collections of the RUSI Museum. The model had been loaned to the Science Museum briefly when the photographs were taken and it was then returned to meet an uncertain fate. The RUSI Museum was since disbanded, and the whereabouts of this lovely model unknown. With these photographs as inspiration, we resolved to track it down. The trail was cold, however, and there was little we could do but hope until one day in 1980 when Henry was rummaging through magazines in a used bookstore. He chanced across a May 1973 copy of *Antique Collecting* magazine, a British publication, and on page 24 was an article on 'Ships' Models'.[1] The first illustration in the article was a broadside photograph of the elusive *Lion*. Henry immediately noticed that this was *not* one of the old Science Museum views, but a more recent image. Most incredibly, the photo credit was for the Parker Gallery. The trail was hot and we immediately phoned Bertram Newbury, the proprietor of the venerable gallery, who was only slightly helpful. He acknowledged handling the model, but would only reveal that he had sold it to a fellow American living abroad. Understandably, he was not at liberty to give out further details. We took him up on his offer to forward a letter to the owner, but no response ever came.

One or two years passed with no further progress, until Henry brought up the subject while visiting Donald MacNarry. Mr MacNarry is one of the world's best modellers and has had a lifelong interest in Admiralty Board models. He pulled a photograph out of his files that had been given to him by Mr Newbury, and it was another black-and-white view of the *Lion*. On the back of the image, however, a series of letters and numbers had been inscribed. It was not a telephone number, and both Donald and Henry were puzzled. Our father, however, suggested that it looked like a USECOM number, a US military ID number used by personnel in Europe. We rushed to the nearest library and spent the next several hours calling every American military base in Germany. In Stuttgart we struck gold. Henry was connected to a civilian legal attaché. After years of dreaming about this moment, Henry could hardly believe that he was actually talking to the owner of this exquisite model. The gentleman confirmed that, yes, he had purchased the model from Mr Newbury, and when asked if he would consider selling it, he

This photograph shows, from left to right, Henry, Roman and Arnold with the *Lion* in 1983.

The main companionway leading to the upper deck is a finely constructed bell stairs, and the balusters of the quarterdeck railing are carved with the same barley twist as on the stern and beakhead, but without the gilding.

informed us that it was in a bank vault for safe keeping because he had moved into smaller quarters several years ago and no longer had room for it. This we took as a good sign. He promised to think about selling it and call Henry at his hotel in London later that evening. Time passed slowly until the phone rang. He began the conversation by saying that he had retrieved the model and was looking at it and assured us that it was in the same condition as when he had bought it. Henry immediately sensed what was coming next, and it remains one of the greatest moments in our collecting career. The owner agreed to sell it, for a specific price, 'Take it or leave it.' Our father, who had just returned to New York from a trip to Switzerland, and Henry were in Stuttgart the next day.

Provenance

This model first came to light in the collections of the Royal United Services Institution Museum, where it was deposited by E E Rushworth in 1911. By 1973 it had been de-accessioned and was sold by the Parker Gallery, London, to a collector, from whom we purchased it in 1978.

The name *Lion* is painted on the centre of the upper counter, which is unusual at this early date. The breast rail of the open gallery is supported by barley twist balusters that reverse hand in the midline to achieve symmetry, an indication of the exacting care taken by the modeller.

THE *LION*, 4TH RATE c1738

We have two copies of this rare model maker's trade card in our collection, and one came attached to the baseboard of the *Lion* model. The baseboard does not appear to be original, and it may be that Mr Hunt cleaned the model rather than built it.

Description

CONDITION

This model has survived in remarkably good condition. When acquired, there was an eighteenth-century model maker's trade card attached to the baseboard, advertising the services of Alan Hunt, who built and also 'kept and cleaned' models in Southwark, London (see Chapter 12). He was active in the second half of the century and would not have made this model. The carved wooden cradles, which are not original, are identical to the ones on a model of the *Royal William* of 1719 at the NMM (model 1719-2, formerly in the Royal Naval Museum). The two models are quite different in style and workmanship, but it is possible that Alan Hunt replaced the original cradles and baseboards of both models later in the eighteenth century.

CONSTRUCTION

Scale: 1/60   Hull length: 34½in

The vessel's name, *Lion*, appears painted on a small label on the upper counter, and the dimensions match the ship as rebuilt in

The lion figurehead has a youthful look. The head is unusually commodious as there are roundhouses (with round windows), seats of ease in front of each roundhouse, and additional ones abreast of the bowsprit complete with discharge pipes. The barley twist balusters above the beak bulkhead, also appearing on the stern, help unify and balance the overall design.

1738. The hull is fitted with orlop deck clamps, beams, and carlings, and the main wales are built up of three flush black strakes with a white plank below that extends around the wing transom and knee of the head. The topsides are planked with separate strakes, as is the anchor lining, and rigols are fitted above each gun deck port not protected by a channel. Inboard details include an elegant bell stair leading from the quarterdeck to the waist, a diagonal chessboard pattern inlaid on the floor of the captain's cabin, and a single wheel in its 'new' position forward of the mizzen mast. Bulkhead screens and doorways feature lights with delicate wooden mullions and ebony inlay on panels and pilasters. There are two pairs of tiny hinged doors on the forecastle bulkhead to be used as gun ports to repel an enemy boarding party.

This model features an unusual amount of gilding on both carved surfaces and mouldings. Gilded edging adorns the cheeks, head rails, beak bulkhead pilasters and channel capping pieces. The delicate barley twist balusters above the beak bulkhead are also gilded, as are the decorations on the catheads. Ebony inlay embellishes the panelling on the poop bulkhead and taffrail.

Carved work includes delicate barley-twist balusters supporting the breast rails at the beakhead, stern gallery, and between the lights of the quarter galleries. Remarkably, the twist of these tiny gilded balusters switches from right to left-handed at the centreline! The monogram of King George II is delicately carved on the upper finishing of the quarter galleries, and his portrait bust appears in the centre of the taffrail.

The painted decoration is in chinoiserie and includes Chinese-style figures painted in black, red, and gold lacquer on the

There is no bulkhead aft on the upper deck and no central gangway on the quarterdeck, though an elaborate bell stair leads to the waist from the forward end of the deck. A single wheel is mounted forward of the mizzen mast. The white strake at the waterline is an unusual feature.

beakhead bulkhead and stern. The frieze is decorated in bands of chevrons, foliage, and trophies of arms, and each gun port lid sports a different version of an oriental anthropomorphic lion's face.

Literature

The following references include photographs and descriptions of this model:

Edson, Merritt, 'Rigging Data for Two British Fourth-Rates', *Nautical Research Journal*, 23, 4 (June, 1977), pp. 90–8.

Ellis, C Hamilton, *Ships* (New York: Peebles Press International Inc, 1974), Figure 84.

Franklin, John, *Navy Board Ship Models 1650–1750* (London: Conway Maritime Press Ltd, 1989), pp. 29, 31, 38, 57, 154–156, colour plate 8.

Hough, Richard, *Fighting Ships* (New York: G P Putnam's Sons, 1969), p. 148.

Hough, Richard, *A History of Fighting Ships* (London: Octopus Books Limited, 1975), p. 59.

Hughes, Eleanor, ed, *Spreading Canvas* (New Haven, Yale University Press, 2016), p. 214.

A carved bust of King George wearing a laurel wreath appears in the centre of the taffrail, flanked by female figures holding crowns. Alongside are cupidons blowing trumpets and bearing shields with a 'G' carved in relief on the port side and an 'R' on the starboard that, taken together, constitute the monogram of King George II.

Kobak, Laurence B, 'British Admiralty Model Collection', *Sea Heritage News*, 4, 12 (1983), p. 6.

Kriegstein, Arnold and Henry, 'The Kriegstein Ship Model Collection', *Nautical Research Journal*, 27, 2 (June, 1981), p. 81, 91.

Kriegstein, Arnold and Henry, 'The Kriegstein Collection of British Navy Board Ship Models', *Nautical Research Journal*, 38, 4 (December 1993), p. 233, plates 17, 18.

Leetham, Arthur, *Official Catalogue of the United Service Museum* (Southwark: J J Keliher & Co, 4th edition, 1914), Additions to the catalogue, p. 20.

Roth, Leah and Ilene, 'Mirror Image', *Motor Boating & Sailing*, 148, 4 (October 1981), p. 36, 37.

Thring, 'Ship's Models', *The Journal of Antique Collecting*, 8, 1 (May 1973), pp. 24–7.

Walker, Grant H, *The Rogers Collection of Dockyard Models*, Vol. 2 (Florence, OR: SeaWatch Books, LLC, 2018), p. 130.

Williams, Guy, R, *The World of Model Ships and Boats* (New York: G P Putnam's Sons, 1971), p. 72.

Exhibitions

London, Royal United Service Museum, 1911–65.
New Haven, Connecticut, Yale Center for British Art, *Spreading Canvas*, 14 September–4 December 2016.

The captain's cabin can be seen here with its inlaid pilasters and geometric flooring. The ensign staff fittings include a pivoting step for lowering the flag and a metal tabernacle for securing it to the taffrail. The upper finishing of the quarter gallery is adorned with a fine monogram of King George II.

Historical Perspective

THE *LION* AND THE YOUNG PRETENDER

The *Lion* was originally built at Chatham dockyard by Benjamin Rosewell and launched on 20 January 1709. In 1738, she was rebuilt at Deptford dockyard as a 58-gun 4th-rate ship under the supervision of Richard Stace, and our model represents the ship at this rebuilding. She served in the West Indies in 1740 and in Cartagena in 1741. The notable events in the history of great warships are not always glorious. As a reflection of the difficulties of manoeuvring large wooden warships in close quarters, the *Lion* was involved in a collision with the *Victory*, 100-gun flagship of Admiral of the Fleet Sir John Norris, off Spithead in 1741. In consequence, the *Victory* lost her head and bowsprit, and the foremast of the *Lion* was carried away along with twenty-eight seamen thrown overboard by the shock.

The Glorious Revolution and exile of King James II was not universally applauded in Britain, and after the union with Scotland in 1707 and death of Queen Anne in 1714, extremists made persistent attempts to restore the Stuart monarchy. In 1715 a Jacobite attempt to place James Edward (the old pretender), son of James II, on the throne failed, but this did not dissuade his son, Charles Edward Stuart (bonnie Prince Charlie), from campaigning for the cause. As it happens, the final Jacobite rebellion, in 1745, was nearly aborted by the *Lion*.

In 1745 most of Britain's army was in Flanders fighting in support of the House of Austria against France and other continental forces. Hoping to divert British attention, France decided to aid the Young Pretender. Charles Stuart and seven followers embarked at St Nazaire on board a small vessel, the *Dentelle*, lent by a sympathetic merchant. He also had arms for about 2,000 men and £2,000 on board, and he sailed on 7 July. His escort, the 64-gun *Elisabeth*, joined off Belle Île with orders to round Ireland to land on the west coast of Scotland.

On 7 July, The *Lion*, captained by Piercy Brett, encountered the small expedition and immediately gave chase. The engagement began at 5 o'clock when the *Lion* poured a broadside into the *Elisabeth* at short range and continued unabated for five hours. By then, the *Lion* had sustained so much damage to her rigging, she could not make sail. For her part, the *Elisabeth* had suffered mainly in the hull, where she was so peppered with shot holes that several of her gun ports had been knocked into one. The *Dentelle*, which had been beaten off by the *Lion*'s stern chasers, crowded-on sail and

Right: Trophies of arms in oriental style embellish the decoration of the frieze planking along with foliage and chevrons in low relief. The open stern gallery extends around the quarters, and the lions on the gun port lids sport anthropomorphic whiskers.

Left: Each of the whiskered lion heads painted on the gun port lids has a different expression. This one seems more pensive than fierce.

THE *LION*, 4TH RATE c1738

escaped, reaching the coast of Lochaber by late July. The *Lion* lost sixty-five men in the combat, with an additional 107 wounded, of whom another seven eventually succumbed to their wounds. French losses were slightly greater with sixty-five killed and 136 wounded.

Once in Scotland, Charles took Edinburgh with an army of 2,400 Highlanders and marched into England, reaching as far as Derby by December. Support from northern English Jacobites failed to materialise, however, and British naval forces cut off any support from French sympathisers. Charles was forced to retreat into Scotland, where his forces were decisively defeated on Culloden Moor on 16 April 1746. With brutal force, the Highlanders were finally brought under central government control. Prince Charlie managed to escape to France, in a harrowing journey during part of which he was disguised as a maid. He died in Rome in 1788.

In 1747, under command of Captain Arthur Scott, the *Lion* took part in the Battle of Cape Finisterre, where British victory led to the peace treaty concluded between Britain and France in April 1748. This treaty failed to deal with the right of British ships to navigate the American seas without being searched and left unsettled the ownership of Nova Scotia, both matters that had been in contention before the war. Hostilities came to a head with the French invasion of Minorca, and war was declared again on 18 May 1756. The *Lion* was in action again in the West Indies where her captain, William Trelawney, was wounded in the capture of Guadeloupe. In the last year of the conflict, 1762, the *Lion* captured the French ships *Zephyr* and *Ecureuil* before hostilities ended without a single fleet engagement. She was sold out of the Navy in 1765.

JAPANNING IN EIGHTEENTH-CENTURY DECORATIVE ARTS AND WARSHIP MODELS

Some of the most beautiful eighteenth-century ship models are those that were decorated in chinoiserie. The influence of Chinese decoration, and in particular Chinese lacquer work, began with the importation of goods from China and Japan by the East India trading companies in the seventeenth century. The impact of lacquer finishes and Chinese decorative motifs can be traced to the publication in 1688 of a book with the unwieldy title of *A treatise on Japaning and Varnishing, being a compleat discovery of those arts … together with above an hundred distinct patterns for Japan-work, in imitation of the Indians, for tables, stands, frames, cabinets, boxes, etc* by John Stalker and George Parker.[2] This book introduced the term 'Japanning' and triggered a domestic movement to imitate the expensive techniques of Eastern lacquer finishes, not only by providing detailed instructions on how to prepare pigments, varnishes, and lacquers, but also by providing samples of chinoiserie in the form of twenty-four engravings of decorative motifs including pagodas, Chinese figures in oriental costume, exotic animals, birds, etc. These patterns were intended for use not only by professionals, but also by amateurs, and a virtual cottage industry of Japanning blossomed among aristocratic women with time on their hands. The authors actually encouraged readers to cut out the pages containing the patterns and instructed them to apply the patterns directly to furniture in order to transfer the designs accurately. Chinoiserie decoration began appearing on furniture, house decoration, tapestries, etc. in remarkable quantities but with varying degrees of success reflecting the uneven skills of the amateur artists. Many examples of furniture decorated this way have survived, and it is remarkable how seventeenth- and eighteenth-century 'Japanned' objects prized by collectors today have decorations copied directly from the patterns illustrated by Stalker and Parker. The popularity of chinoiserie reached a peak in the eighteenth century and extended to nearly every aspect of the decorative arts including, remarkably, the decoration of warships, or at least, models of some of them, as reflected by the lacquered decoration in the chinoiserie style that adorns this exquisite model of the *Lion*.

References

Clowes, Wm Laird, *The Royal Navy a History From the Earliest Times to the Present*, Vol. 3, (London: Sampson Low, Marston and Company Ltd, 1898).

Right: This is the second model in our collection to be decorated in chinoiserie style. In addition to the gilded carving, there is abundant lacquer work featuring gold relief against a black background, with red touches combining to create a rich colour scheme. The forecastle bulkhead is elaborately constructed with ebony inlay and glazed lights for the boatswain and carpenter's cabins, and the double doors providing access to the forecastle contain small, divided gun port lids to enable a gun to be run out facing aft to repel boarders.

THE *LION*, 4TH RATE c1733

This model has suffered losses over its life but has had only minor repairs with no replacement of missing parts. The fully planked hull is typical of dockyard models built after the middle decades of the eighteenth century.

CHAPTER 13
A George II 4th rate c1745

~ Acquisition ~

THE SOMBRE COLOURS AND lack of decorative detail underscore the simple beauty of sheer and tumblehome in this finely made hull model. Because models such as this were built at the dockyards by men engaged in shipbuilding, they may embody secrets of design and construction that have yet to be discovered. This model was purchased at auction in Maine in November 2002.

~ Provenance ~

This model was bought at auction and, as is too often the case in such circumstances, its prior history is unfortunately not known.

~ Description ~

CONDITION

This model has lost most of its fittings and decoration, and when acquired there was some obvious 'restoration', all of which we removed. In its present state, there are no replacements and the model, though incomplete, is entirely original.

CONSTRUCTION

Scale: 1/48 Hull length: 40in

This model is of an unusual and distinctive style, and is built of mahogany, boxwood, horn and ebony. It would, in fact, be unique were it not for one other example that was sold at auction on 17 November 1971 by Christie's in London.[1] This was of a 76-gun ship, but the materials and construction were identical and it is most likely by the same builder.

The keel and stem are connected by an unusual interlocked scarph, and the hull is framed and finely planked with mahogany strakes held in place by brass pins. The wales are of horn and are pierced at the level of the floor of the gun deck to accommodate eight lead scuppers on each side. Lead hawse pipes are also fitted, and there is an ebonised beakhead bulkhead. The gun deck is planked, and internal details include ridding bitts and a serrated rack for the jeer capstan.

~ Historical Perspective ~

THE ESTABLISHMENT STIFLES INNOVATION

This model conforms to the 1745 Establishment for 4th rates of 50 guns, of which there were seven launched between 1747 and 1757. The Establishment of 1745 was one in a series of attempts to

The lower counter is fully planked, but the missing stern reveals the stern timbers and deck transoms. Markings indicate the outline of the quarter galleries, now missing.

standardise the construction of men-of-war in order to gain efficiencies in the fitting and repair of naval ships. The first 'Establishment' was a set of guidelines for the dimensions of line-of-battle ships that appeared in 1706 and came to be known as the '1706 Establishment'. Under the guidance of the Surveyor of the Navy, subsequent codified instructions were issued in 1719, 1733 and 1741, each tending to specify stricter dimensions, until the Establishment of 1745, which went so far as to dictate the actual lines to be used for each class of ship. Belying their original purpose, none of these Establishments lasted long enough to generate a uniform Navy, and they stifled innovation and resulted in a fleet full of anachronisms and imperfections. For example, by 1745, the old 1st and 2nd rates, originally intended for large fleet engagements, were too large and cumbersome for naval tactics that

All planking is held in place by brass pins, and the foremost scuppers, made of lead, can be seen at the level of the main gun deck.

had evolved to favour amphibious operations in shallow coastal waters and in harbours that often could not accommodate the draught of a 'Great Ship'. Larger two-deckers were already favoured by Continental navies, and the British gradually came to appreciate their advantages.

The 50-gun ships of the 1745 Establishment were larger than their predecessors and enjoyed some success against the French and the American colonists, but were eventually squeezed out by the more powerful 74s and the more manoeuvrable frigates of the latter part of the eighteenth century.

A GEORGE II 4TH RATE c 1745

Left: The coamings for the hatches and ventilation gratings can be seen on the quarterdeck. The waterway on the poop deck, and adjacent plank on the quarterdeck, are of darker wood than the rest of the decking.

Below: The planksheer is missing, revealing the frames. The upper decks are planked in a lighter wood than the main deck, which is best appreciated when viewed from above.

This is one of the first two-deckers to have fore and main channels positioned above the upper gun deck. The model maker has taken care to block each gun port with glass backed by a metallic fabric, coloured red where lids are present and blue for upper deck ports lacking lids.

CHAPTER 14

The *Namur*, 3rd rate of 1746

Acquisition

THE INTRINSIC BEAUTY OF a finely made ship does not depend on painted work or carving, but derives from a marriage of form and function to produce a hydrodynamic shape that can swim even in the mind's eye. Despite its great size, this impressive 74 has the graceful lines of an aquatic bird, and the *Namur* model seems to float in the water even when sitting on its case. This model was offered for sale at auction by Sotheby's Park Bernet on 30 May 1974. It failed to sell, and the next day we made an offer to the owner, via Sotheby's, which was accepted one week later.

Provenance

This model was acquired by Junius S Morgan Jr in the first half of the twentieth century. Its earlier history is unknown.

Description

CONDITION

While owned by Junius Morgan Jr, this model was suspended over a doorway without protection. When we acquired it, there were some losses consisting of bits of carving from the taffrail and quarter galleries, some mouldings, and a few gun port lids. These were replaced by August Crabtree of Newport News, Virginia, in 1975.

CONSTRUCTION

Scale: 1/48 Hull length: 48¼in

The model has a highly distinctive taffrail depicting a turreted castle in the centre, flanked by effigies of Hercules performing two of his twelve labours. On the starboard, Hercules is depicted with the Nemean lion and to port, with one of the Stymphalian birds. Shortly after acquiring the model, Arnold was visiting the Tate Gallery in London and discovered that the painting by Samuel Scott of the Battle of Cape Finisterre, fought in May 1747, depicted a ship with exactly the same stern as on our model, engaging a French two-decker in the left of the painting. A description of the painted scene was provided by the museum administration, and it claimed that the ship to the left was the *Devonshire*, a third rate of 66 guns built in 1745. For many years we assumed this was the identity of the model, but the significance of the castle on the taffrail remained a mystery. One day, while re-examining the list of participants in the battle, Arnold discovered that the *Namur* (74) captained by the Honourable Edward Boscawen was in the thick of the fight, and that Boscawen was severely wounded, though not fatally. The important

The trailboard carving features a triton blowing his horn and behind him a dolphin nestles against the cheeks.

role played by the *Namur* in the battle would justify a prominent place in Scott's magnificent painting, possibly more so than would the *Devonshire*. Most importantly, the *Namur* was named for the Belgian town whose greatest monument is its ancient citadel. This finally explained the significance of the castle on the taffrail. In addition, the identification fits the model perfectly. The *Namur* was rated a 74, as is the model, but the *Devonshire* was rated a 66, a possible rating for the model, but really not a very good fit.

The position of the fore and main channels on the model are raised above the main deck gun ports, a shift that was instituted in 1745, making this ship, built in 1746, one of the earliest vessels

This block model features a finely formed and hollowed out hull with score marks to indicate planking. There is a solid wale, and the upper surfaces of the model are covered in sheets of wood to which varnished paper has been glued.

to reflect this new feature. The mizzen channels are also raised and mounted on the sheer rail above the quarterdeck gun ports, and these innovations, along with the new number and arrangement of the gun ports, may have justified the construction of this elaborate block model.

The hull is carved from solid wood with great precision, and the seams between the planks are represented by deep score marks.

The identification of this model is confirmed by the decoration of the taffrail. Featured centrally is the ancient citadel of Namur, and flanking this are effigies of Hercules performing two of his labours.

All of the gun port openings and stern lights are closed with a thin sheet of glass backed by metallic fabric, concealing the interior of the hull. The fabric employed is coloured red behind the gun port lids but is blue behind the ports on the upper deck in the waist, which are not fitted with lids. The top of the model is completely covered by wooden sheets at the level of the plank sheer, and the wood, in turn, is covered by varnished paper that bears the London watermark of a mid-eighteenth-century type made by Lubertus van Gerrevink. There are metal collars provided to support poles for launching flags, including a hole in the lace for a jack, but the poles and flags are unfortunately missing.

The exterior of the model is fully decorated with painted Trophies of Arms on the beak bulkhead, upper counter and topside frieze aft. At the head, there are two-tiered seats of ease in addition to roundhouses.

Literature

The following references include photographs and descriptions of this model:

Brown, C, 'Down to the Sea in Ship Models', *Forbes*, 144, 11 (13 November 1989), pp. 336–40.

Franklin, John, *Navy Board Ship Models 1650–1750* (London: Conway Maritime Press Ltd, 1989), pp. 28, 29, 30, 174–175.

Kobak, Laurence B, 'British Admiralty Model Collection', *Sea Heritage News*, 4, 12 (1983), p. 6.

Roth, Leah and Ilene, 'Mirror Image', *Motor Boating & Sailing*, 148, 4 (October 1981), p. 37.

Kriegstein, Arnold and Henry, 'The Kriegstein Ship Model Collection', *Nautical Research Journal*, 27, 2 (June 1981), p. 92.

Kriegstein, Arnold and Henry, 'The Kriegstein Collection of British Navy Board Ship Models', *Nautical Research Journal*, 38, 4 (December 1993), pp. 223–4, plates 19, 20.

Historical Perspective

The *Namur* was launched after the French had joined the Spanish against the British in the War of Jenkins's Ear, begun in 1739. The French navy had adopted the 74-gun ship as a key element in its naval force, and the British had no equivalent weapon. The *Namur*, a 90-gun ship of the 1719 establishment, was cut down by one deck to create the first 74-gun ship in the British Navy. Launched in 1746, she became the first in a long pedigree of warships that formed the backbone of the British battle fleet until the end of the sailing warship era. Although a step in the right direction, she was smaller and lighter than her French contemporaries, and it was not until captured French 74s were available later in the war that British designers adopted their heavier lines.

THE *NAMUR* AND THE WAR OF JENKINS'S EAR

Britain and Spain each had colonies in the New World by the end of the seventeenth century, and in the eighteenth their mercantile interests clashed. The flashpoint was in the West Indies, where the Spanish had for years captured and plundered British vessels. The most infamous incident allegedly occurred when Richard Jenkins, master of a brig sailing from Glasgow, was boarded by a Spanish vessel out of Havana on 20 April 1731.[1] The Spanish captain reputedly cut off one of Jenkins's ears and handed it to him saying: 'Carry this home to the King, your master, whom, if he were present, I would serve in like fashion.' This account was given by Jenkins at a hearing before the House of Commons in 1738, during which he responded to a question regarding his feelings at the time of his mutilation with the rousing answer: 'I recommended my soul to God, and my cause to my country.'

On 23 October 1739, war was declared against Spain. France subsequently joined with Spain, and the Dutch joined the British.

In 1747 the *Namur*, commanded by the Honourable Edward Boscawen, was attached to a British squadron under Vice Admiral George Anson and Rear-Admiral Peter Warren. They were dispatched to intercept a French convoy, which they sighted and engaged off Cape Finisterre on 3 May. The French ships were under protection of a squadron commanded by the Marquis de La Jonquie're, who faced the British with twelve of his best ships. A running fight ensued from 4 to 7pm, when the French struck their colours. Eight French ships were captured, making this one of the most decisive victories of the first half of the eighteenth century. In the action, Captain Boscawen was wounded, but he recovered quickly and later in the year was sent to the West Indies as Commander-in-chief on board the *Namur*.

In 1748, Boscawen took the *Namur* to the Indian Ocean, where she participated in the unsuccessful siege of Pondicherry. In April of 1749, after the war with France had ended, the *Namur* was detached to assist the East India Company in a war with the King of Tanjore. While doing so, she was caught in a violent hurricane and wrecked with the loss of 520 men. Thus the *Namur* enjoyed a brief but eventful life at sea, and as a prototype 74, she played an equally important part in the evolution of British warship design.

The beakhead is equipped with roundhouses and seats of ease, and the head rails though of simple form, are decorated. The figurehead is a youthful lion.

Right: There is considerable tumblehome that is best appreciated when viewed head-on.

THE *NAMUR*, 3RD RATE OF 1746

CHAPTER 15

A French 64-gun ship c1754, built by Augustin Pic

The archives of the Bibliotheque National in Paris contain a letter written at Rochefort Dockyard dated 4 August 1759. The letter, which appears to be a critical review of the state of the dockyard, reports on a series of activities considered detrimental to the operation of the dockyard. Interestingly, one reported concern was that 'everyone wants to have a model of a warship' and as a result 'the best workers are engaged in this expensive activity'.[1] Augustin Pic was a thirty-two-year-old Student Naval Constructor at Rochefort dockyard when he made the model described in this chapter in 1754. He was engaged in making another model a year later, and made his last-known model in 1757. Pic was almost certainly one of the offending workers cited in the critique, and the timing suggests that subsequent repercussions may have led to the untimely end of his modelling career.

Acquisition

This model was originally offered to us in 1988 by Mr J P Dieutegard, a Parisian maritime antiques dealer. Henry went to Paris to see the model and, if merited, to try and negotiate a purchase. The model was in Mr Dieutegard's apartment, and while inspecting it, Henry decided to make an offer. But when they retired to the study, after a glass of wine, Mr Dieutegard brought up the subject of the model with the following comment: 'If you are thinking of offering … [and he stated the precise price that Henry had in mind] don't bother, because at that price I am buying, not selling.' Well this took the wind out of Henry's sails, who had no choice but to 'not bother' to make the offer. The result was that the model was acquired by Mr and Mrs Larquetoux for their private museum in the Vauban Citadel on Belle Île. But in 2010 the Citadel was sold to a hotel chain and the contents of the museum were dispersed at auction on the premises. Arnold had the pleasant task of attending the sale, which included a very memorable overnight stay at the Citadelle Vauban Hotel, and this time we acquired the model.

Provenance

Details concerning the creation of this model and its early provenance have been preserved thanks to a curious document held at the National Archives in Paris entitled: 'Extrait des Services du Sieur Augustin Pic'.[2] Augustine Pic held the position of student naval constructor at Rochefort dockyard in 1752 when he began building this model of a 64-gun ship. The model was completed in 1754 and was presented by Pic to Sébastien-François-Ange Le Normant de Mézy, the Intendant of the Rochefort dockyard. Le Normant de Mézy had become the Intendant of Rochefort in 1750, and he most likely commissioned the model as Pic was looking for

Left: Starboard and port profiles. On the starboard side alternating strakes of planking have been omitted beneath the waterline, permitting a view of the bends (consisting of paired frames), inboard strakes of planking, etc. The port side is fully planked. Note that only the ports of the main deck battery have lids, as was customary.

A vignette chapter heading after Ozanne taken from Duhamel Du Monceau *Elemens de L'Architecture Navale*, 1752. The image shows a room of the Academy of the Navy at Brest with ship models on display.

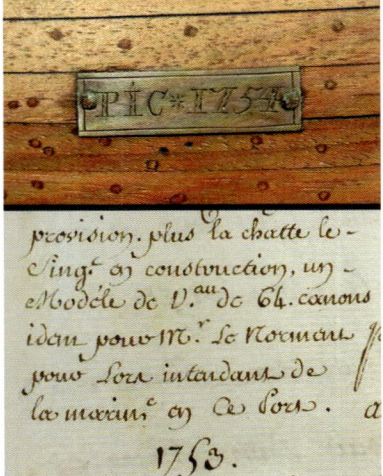

Top panel: The small silver plate nailed to the starboard cheek of the Pic 64-gun ship engraved with Pic's name and the date 1754. Note that the strakes of planking are held in place by hundreds of tiny wooden trenails. Bottom panel: Notation in the margin of Pic's Service Record stating that he was building a model of a 64-gun ship for Mr Le Normant, Intendant of the naval dockyard at Rochefort in 1753.

Part of a letter written at Rochefort Dockyard and dated 4 August 1759 reporting that 'everyone wants to have a model of a warship' and as a result 'the best workers are engaged in this expensive activity'. Pic was making models at Rochefort Dockyard from 1752 to 1757, so that it is more than likely he was one of the 'best workers' referred to in this note. However, the note also suggests that Pic was not the only one building ship models at Rochefort. Other models may eventually be identified that were also constructed at the Rochefort arsenal at this time. (Author's photograph, courtesy of the National Archives, Paris)

The female figurehead has a strikingly asymmetric pose, with one arm poised behind her head and her head turned to face starboard. The main rail of the head terminates in an elaborate carving that incorporates the cathead knee and connects to another free-form decorative terminus of the middle-rail that extends on to the planking alongside the rail. Note the knight's head ends on the stanchions at the forward end of the forecastle and the sheaves in each stanchion. The triangular wash cant is clearly seen beneath the lower cheek, and defenders have been fitted between the cheeks to protect the gammonings. The outboard profile of the five pairs of head timbers bear elegantly moulded covering boards. The lead linings of the hawse holes are also visible. Note the complex curves of the boomkins that can be seen fayed to the sloped gratings of the head. The catheads are quite detailed, with iron bands at their ends and three sheaves turning on an iron arbor secured with cotterpins. The fore jeer capstan and the galley hatch with its scuttles for the chimneys are abaft the fore topsail sheet bitts. And note the 'barbette' ports for the forecastle guns. The decorative outline of the anchor lining can be seen aft of the first gun port on the main deck, between the upper and lower wales. The metal knees on the upper side of the fore channel can be seen. Note how they lap over the sheer rail.

ways to supplement his income at the time. Mr Le Normant was the first owner of the model, but it subsequently disappeared for about 100 years until resurfacing as a bequest to the Smith-Champion Foundation in Nogent-sur-Marne in the last century. It was sold by that institution and was in the possession of the Parisian maritime antiques dealer Mr Dieutegard by 1988. It was sold within a year to Mr and Mrs Larquetoux for their private museum in the Vauban citadel on Belle Île, France. We bought it at auction in 2010.

Description

CONDITION

Before beginning restoration, we carefully considered the state of the model. It appeared to have been altered about fifty years after it was constructed, with the addition of new rigging, associated fittings and an updated stern. At some point after it arrived at Belle Île the anachronistic stern was removed. Henry had examined and photographed the model back in 1988 when it was first offered for sale. At that time the replacement stern, rigging, etc. were still on the model. Detailed inspection led us to believe that the rigging and other additions to the model were of the same age, and based on their characteristics, were probably fitted in an attempt to update the model around the year 1800. We found no evidence that the

The bulbous form of the bow and tumblehome of the hull can be appreciated in this view. Note the omitted strakes on the starboard hull. Other details that can be seen in this view include the doorways leading to the bulkhead, the bow chase ports, and the way that the planking extends forward beyond the cathead supporters to finish at the outer edge of the ladders that lead to the foredeck.

model had been accidentally damaged and repaired or restored; rather, the model was deliberately altered. These alterations included the fitting of an anachronistic stern and quarter galleries and raising of the bulwarks. By the time it was sold by the Citadel museum, the curators had already saved us the trouble of removing the incorrect stern elements.

Deciding how to proceed with the restoration turned out to be rather straightforward. In many places the interface between the rigging and the hull was rather crudely done. Racks had been added for belaying pins, the forecastle and quarterdeck gun barrels and rigging bits had been covered by box-like enclosures, and other crudely made fittings and railings were added to which rigging lines were fastened. New chandeliers had also been fitted for hammock netting, leaving the original supports empty.

We decided to remove all these remaining additions, including the masts, spars and rigging, in order to restore the model to its 1754 appearance. The work was entrusted to Philip Wride. When he removed the heightened bulwarks from the quarterdeck and fore deck, it was gratifying to see that many of the original timber heads, kevels and cleats had been preserved underneath.

The most challenging part of the restoration was creating the new stern carvings. We instructed Philip to base the shape and designs of the carvings on the 74-gun model also by Augustin Pic preserved at the Musée de la Marine. We assumed the original sterns would have resembled each other since, in so many respects, these two Pic models are very alike.

Philip also retouched the gold leaf on the wax figurehead. Aside from the stern, the model had survived in remarkable condition. All sixty-four original guns are preserved, with their tackles and paint. The model retains its original surface, and all the carvings forward of the stern including the breast rails, cathead decorations and the figurehead, are complete and original. There is a wealth of detail barely visible below decks, including a messenger cable, rigged tiller, capstans, ladders, knees, etc.

The fore jeer capstan has mortices for capstan bars and notches to take the pawls. Just abaft the casing of the galley hearth is the hatch for the crew's ladderway with its double flight of stairs. A winged and crowned shield bearing a single fleur-de-lis dominates the forecastle breastwork with no accommodation for a belfry.

CONSTRUCTION

Approximate scale: 1:42 Hull length: 52in

The wood appears to be the same as that used by Pic on his 74-gun ship model, which was primarily pear wood and walnut, although on the 64-gun model it retains a darker patina most likely because the finish on the 74 has been lightened during cleaning. The hull is built in frame and planked. The port side is fully planked, and the strakes of planking are held in place by hundreds of tiny wooden trenails. On the starboard side alternating strakes of planking have been omitted beneath the waterline, permitting a view of the bends (consisting of paired frames) and inboard strakes of planking.

The decoration of this model reflects the elegance and refinement typical of French eighteenth-century practice. The female figurehead has a strikingly asymmetric pose, with one arm poised behind her head and her head turned to face starboard. A similar turn of the head to starboard appears on the figurehead of Pic's 74-gun ship. Gilding has been used to highlight the figurehead, cathead brackets, hancing pieces, and a winged and crowned shield bearing a single fleur-de-lis on the forecastle breastwork. The relatively subdued gilding appears to reflect full-

The ship's side just aft of the main channel. The channels are supported by iron knees underneath, and by iron brackets nailed into the ship's side on top, as shown here. Just above the aft end of the channel is a bracket with a ring bolt to spread its sheet away from the bulwarks. Small, round lead-lined scuppers can be seen that connect to the upper deck waterways. Careful inspection of the rails above the channel will reveal two metal fastenings that would have originally supported stanchions along the waist. These fittings appear throughout the model but none of the original stanchions survive.

size practice based on the depiction of French ships in contemporary paintings. The main rail of the head terminates in an elaborate carving that incorporates the cathead knee and connects to another free-form decorative terminus of the middle-rail that extends on to the planking alongside the rail.

The model was built to closely reflect the construction details of a real 64-gun ship of the time. The triangular wash-cant is clearly seen beneath the lower cheek, and defenders have been fitted between the cheeks to protect the gammonings. The outboard profile of the five pairs of head timbers bear elegantly moulded covering boards. Stanchions end with knight's heads at the forward end of the forecastle, with sheaves inset into each stanchion. The hawse holes are lined with lead as in full-size practice, and small, round, lead-lined scuppers connect to the upper deck waterways with two rectangular lead scuppers draining the dales from the port main pumps. A prominent fore jeer capstan sits on the fore deck, and the galley hatch with its scuttles for the chimneys are visible abaft the fore topsail-sheet bitts. As on his 74-gun model, Pic has made no accommodation for a belfry.

All 64 cannon are present and are housed with their original tackles and lashings. Miniature 'lead' aprons are fitted over the vents

A chesstree is visible bolted to the side just abaft of the fore channel. Linked iron chain plates are visible. The gun port lids each have a pair of wooden toggles spliced to the spans of the port-tackles to allow the ports to be sloped at 45 degrees. Lead-lined scupper outlets are visible to drain the upper gun deck.

to keep them dry and to prevent accidental firing. All the guns are run out with their breech tackles rigged. The single wheel is mounted on metal pillars just beneath the poop. The officer accommodations on the quarterdeck are displayed by a series of small cabins. Open doorways lead to two sleeping cabins, starboard and port, for officers.

Less visible details are present below the waterline, and inboard details include the steps of the masts in the hold, along with the riders, sleepers, strakes of planking, etc. At the orlop deck there are steps let into the sternson knee to provide access to the gunner's stores, and the tiller is rigged to the wheel. Hanging and lodging knees are held in place with wooden trenails, and included on the gun decks are the bitts, capstans, companionways, cabins, etc.

A view of the ship's side with the break of the quarterdeck marked by a gilded hancing piece. The outboard ladder has a rope handhold threaded through the treads. Just forward of the ladder a billet with sheaves for the foresheet and the lower fore studdingsail tack is set into the bulwark. Also visible are two rectangular lead scuppers draining the dales from the port main pumps. At the very top of the photograph can be seen the gangways formed by gratings that link the forecastle to the quarterdeck, supported by pillars.

~ Literature ~

The following references include photographs and descriptions of this model:

Kriegstein, Henry and Kriegstein, Arnold, 'Le Modàle, construit par Augustin Pic en 1754, d'un vaisseau de 64 canons', *Neptunia*, No. 284, December 2016, pp. 20–33.

Boudriot, Jean, 'A Model Builder of the Eighteenth Century', *Neptunia*, No. 130, 1978, pp. 7–20.

Exhibitions

Nogent sur Marne, France, The Smith-Champion Foundation, prior to 1988.
Belle Île, France, Museum of the Citadelle of Vauban, 1989–2010.

A view of the quarterdeck just abaft of the mainmast. All 64 cannon are present and are housed with their original tackles and lashings, including these 6pdrs on the quarterdeck. Miniature 'lead' aprons are fitted over the vents to prevent accidental firing. All the guns are run out with their breech tackles rigged. Despite their small size, all the trucks are decorated with a quarter-round moulding around their external rims. A staghorn cleat is visible, used for belaying sheet and tack lines. Gratings are removed to expose the upper barrel of the double main capstan on the gun deck. It is equipped with eight capstan bars. The bars are octagonal with their heels squared off to insert into bar holes in the drumhead. All the bars are served. The quarterdeck ladder leading to the upper deck is visible, and aft of it on the upper deck are the partners of the mizzen mast. Between the quarterdeck guns the waterways are pierced with round lead-lined scuppers.

The upper deck is shown here, just before the mainmast, with its gratings removed. A ladderway with double flights of stairs is visible. The stringers are lodged against cleats on the gun deck. The rope messenger is visible on the gun deck with its mouses. The partners for the mainmast are missing, but the accommodation for the mainmast as well as the four pumps are visible.

Left: A view of the quarterdeck showing the poop breastwork carved with an openwork frieze, mirroring the forecastle breastwork carving. The single wheel is mounted on metal pillars just beneath the poop. Open doorways lead to two starboard sleeping cabins for officers. Note the moulded corner stanchions of the bulkheads. Ladders providing access to the poop deck are fitted on each side.

A FRENCH 64-GUN SHIP c1754, BUILT BY AUGUSTIN PIC

Above: A view from the quarterdeck looking aft at the wheel and beyond through the passageway between sleeping cabins into the great cabin. This photograph was taken in 2010 after the stern had been removed.

Below: A photograph of the upper gun deck with its battery of 18pdrs taken in 2010 before a new stern made this view impossible. Amidships can be seen the tiller-ropes abaft the mizzen mast, and forward of these an elegant ladderway and then the upper barrel of the main capstan. Forelocked through the binding strakes that run fore and aft is a row of ringbolts for the train-tackles of the guns.

Above: The open planking on the starboard stern reveals the sternpost knee lodged against the sternpost. The rising wood is deeply scored to accept the heels of the floor timbers, and the gudgeon and pintle straps are decorated with spade-shape ends.

Below A hanging knee from the interior of the model, which was found loose. Note the five tiny wooden trenails that Pic set into the vertical arm of this fitting. The ends of each beam on this model are secured to the ship's side with one of these knees, and the care Pic took to fashion each one in this way is quite impressive, particularly when one considers that under normal circumstances this work is entirely invisible.

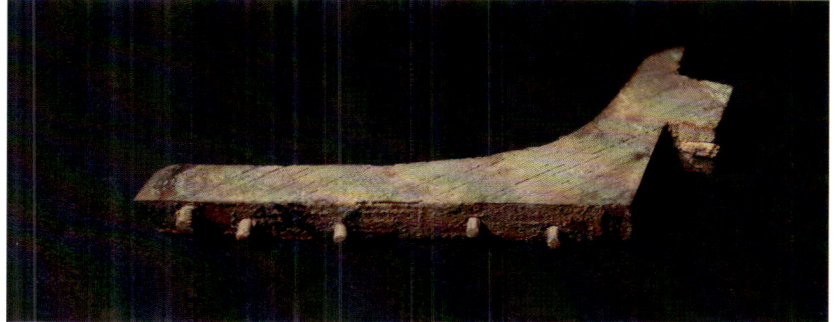

Historical Perspective

AUGUSTIN PIC, SHIPWRIGHT AND MODEL BUILDER

Our collection of seventeenth- and eighteenth-century ship models consists primarily of British dockyard examples, but we do have this French model of a 64-gun ship completed in 1754 at the Rochefort arsenal. The builder was a thirty-two-year-old student naval constructor named Augustin Pic. The vast majority of period models built in the eighteenth century are mysteries as far as their makers are concerned, so it is a near miracle that we know the circumstances of this example's construction. This is due to the custom Pic had of signing and dating his models. He made three miniature French ships: one galley and two ships of the line. Each bears a silver plate engraved with his name and the year of

Overall views comparing the 74-gun model on the left with the 64-gun model on the right. The 64-gun model is half the size of the other, but they are strikingly similar with notable differences. Similarities include the hatchways with gratings that run along the gangboards of both models, the absence of belfries and the omission of strakes of planking to reveal framing details. Interestingly, breastworks are preserved on the forecastle, quarterdeck and poop of the 64, but are not present on the 74. Possibly they were originally fitted on the 74-gun model and have been lost over time. Other differences include the colour and finish of the wood (probably due to restoration on the 74), and small differences in the shapes of some features. It should be noted that the small deck cabins inboard of the taffrail on the poop deck of the 64 are replacements. Photograph of the Pic 74-gun ship courtesy of Musée National de la Marine/P Dantec.

completion of the model. Two are in the collections of the Musée de la Marine, and the other is the subject of this chapter. Personnel records of Mr Pic's service at Rochefort are preserved in a memorandum written by him in 1770–76, so that we also know a good deal about the context of their creation.

From his memoir we can appreciate key events in Pic's career. Augustin Pic grew up in Marseille, a city that housed the Arsenal des Galères, which was one of the largest enterprises of its kind in Europe in the early eighteenth century. The dockyard, although already in decline at the time, is said to have employed around 3,000 skilled and unskilled labourers involved in manning, maintaining and building a fleet of galleys. There is no clear evidence that Pic was actually employed at the dockyard, but Pic's Record of Services states that in 1741–46 he made a model of a galley. The model is currently in the collections of the Musée de la Marine, Paris, and bears the name 'Minerve', although there is no record of an actual galley with that name. This model is the first to carry the characteristic maker's mark that Pic affixed to all his models; in this case the miniature silver plaque is engraved with 'PIC 1746'. Pic tells us that Don Olivares, a general officer in the Spanish Navy, admired the model and offered Pic a significant sum to buy it, but Pic turned him down. Pic kept the model, and it was still in his possession when he penned his Record of Services twenty years later.

The top timbers of the beakhead frame are shaped to form a ladderway to the beakhead. At the top of the image you can just see the way the inboard arm of the cathead curves so that it would not interfere with the recoil of the chase gun. The heel of the boomkins terminate at the knightheads and do not reach the beakhead bulkhead.

Pic became a shipwright through an unconventional route, facilitated by the intervention of the Marshal of Belle Île, with whom Pic was acquainted. Apparently, the Marshal arranged for Pic to apply for appointment as a student naval constructor. He was examined for this appointment in Paris in 1751 by Duhamel du Monceau, Inspector General of the Navy. Pic passed his examination and was appointed student naval constructor at Rochefort dockyard. This position paid only 600 livres, and we are told that Pic had to take on outside work in order to make ends meet.

Aged twenty-nine, he was over-age for such a junior position, and throughout his career Pic was frequently passed over for promotion and grew increasingly frustrated. Pic wrote his memorandum in 1776, and in it he catalogues the slow, unsteady progress of his career as a shipwright. It is a melancholy record of delayed promotions, frustrated building projects and unfulfilled aspiration. One wonders at the lack of recognition by his superiors given the talent and craftsmanship apparent in his models, as well as his knowledge of ship construction reflected by the wealth of accurate detail they contain.

From the Record of Services we learn that in addition to the galley mentioned above, Pic made three other models over a period of five years. In 1752 he began the model of a 64-gun ship that is the subject of the present chapter, which he completed in 1754. Possibly as a result of the success of this commission, he was hired in 1753 to make a model of a 74-gun ship, at a larger scale, for the salon of the Academy of the Navy at Brest. This model was completed in 1755 and is currently on display at the Musée de la Marine, Paris. In 1756–57 Pic made his final model, this time of an English ship, the 60-gun *Warwick*, which he made in the 'English style'. This model was possibly made on his own initiative as he does not state why or for whom it was made. The *Warwick* had been captured by the French in 1756 and Pic was working on her refit at Rochefort in 1757. With the actual ship available, he was in an excellent position to make an accurate miniature. Regrettably, the fate of this model is unknown.

In January 1758, at the age of thirty-six, Pic was finally appointed junior constructor. It is possible that Pic's predilection for building ship models had an adverse effect on his career. Pic did not advance to junior engineer constructor until 1765, and he finally became an engineer constructor in 1767. Throughout his career he did not design or build any major vessels, and he retired in 1780 at the age of fifty-eight. Thus Pic's legacy as a shipwright consists of the models he made rather than the actual ships he aspired to build.

HISTORICAL CONTEXT OF THE MODEL

In the French navy, ships carrying 60–64 guns and measuring generally less than 140ft were classified as '2nd rate, second order', from 1689 until 1734. They were usually pierced for twelve lower-deck gun ports. In 1734 the navy introduced a formula for 64-gun ships that increased their armament to thirteen lower deck ports and increased their length to 150ft. The cannon that these '3rd rate, first order' 64s carried were 26 24pdrs on the lower deck, 28 12pdrs on the upper deck and 10 6pdrs on the quarterdeck and forecastle. This is the class of two-decker to which Pic's model belongs. The last of these warships was laid down in 1779, as the role of these light two-deckers was supplanted by the more powerful 74s on the one hand and the lighter and more agile 12pdr frigates introduced in the 1740s on the other. Part of the importance of Pic's model is that it dates from the heyday of the 64-gun ship, when these light two-deckers formed an integral part of the French line of battle.

The arsenal at Rochefort turned out only one 3rd rate, first order 64-gun ship after the codification of 1734, and it was not laid down until 1748. Then, in 1752, an unprecedented three keels were laid down for 64s! We feel it is no coincidence that the Navy Supervisor commissioned Pic to produce his model of a 64-gun ship that same year. He worked on it until 1754, exactly in parallel with the arsenal's construction of the three ships, viz. the *Inflexible*, *Capricieux* and *Éveillé*. We have not been able to identify Pic's model as a specific ship, but we feel it must be a good representation of this group.

It is interesting to note that Pic made each of his models at a time when he had a ship of exactly the same size and rate at hand. He made his galley model while associated with the Arsenal des Galères in Marseille; his 64-gun model while three 64s were being built at Rochefort; his 74-gun model while he was assisting in the construction of the 74-gun *Glorieux*; and the *Warwick* while he was assisting in the refit of the captured ship. His models have outlived all of their archetypes and are unique time capsules that accurately reflect the construction practices of the period.

Few arsenal models of French 64-gun ships survive. There are three at the Musée de la Marine, Paris: one representing *L'Assuré* of 1740: one of the *Vengeur* of 1756: and one of *L'Artésien* of 1762. A model of the *Prince of Parma* of around 1747 is preserved at Il Teatro Farnese di Parma in Parma, Italy, and there is a model of the *Bien-Aimé* representing a 64-gun ship of around 1740 at the University of Bologna in Bologna, Italy.

References

Boudriot, Jean, 'A General Discussion on French 64-Gun Ships of the Line', *Nautical Research Journal*, Vol. 28, No. 4, 1982, pp. 185–194.

Boudriot, Jean, *Historic Ship Models*, A.N.C.R.E., Nice, France, 1997.

Boudriot, Jean, 'A Ship Model With Coach-House', *Nautical Research Journal*, Vol. 28, No. 2, 1982, pp. 63–78.

Boudriot, Jean, 'The 64-Gun Ship *Artesien* 1762–1785', *Nautical Research Journal*, Vol. 29, No.1, 1983, pp. 35–46.

Boudriot, Jean, & Berti, Hubert, *Les Vaisseaux de 50 et 64 Canons 1650–1780*, A.N.C.R.E., Nice, France, 1994.

Right: The original stern and quarter gallery carvings had been removed long before we acquired the model. Philip Wride carved the new stern seen here to resemble the one on Pic's 74-gun model, which was completed just two years later.

A FRENCH 64-GUN SHIP c1754, BUILT BY AUGUSTIN PIC

This model is incomplete for two reasons. Firstly, there is no evidence that it was ever fitted with a figurehead, head rails, channels, lids for the broadside guns, skids or quarter galleries. Secondly, it suffered from loss of the upper works above the beams for the forecastle and quarterdecks. We elected to reconstruct the profile of the missing topsides at the stern but have replaced no other missing parts.

CHAPTER 16

The *Généreux*, 3rd rate of 1785

Acquisition

MUCH OF THE INTEREST in this unusual model derives from the remarkable workmanship and detail hidden below the decks. We purchased this model from the Parker Gallery, London, in 1990.

Provenance

This model was sold to a London dealer after the Second World War, and its earlier history is sadly unknown.

Description

CONDITION

This model was never fitted with gun port lids, channels, quarter galleries or decorative carving. Whether this model was never finished or designed only to show internal details is not known. When acquired, the glue used to attach the fine planking and hull strakes had dried and come loose in many places, and the topsides above the upper deck spirketting were largely missing. Loose pieces were carefully removed, cleaned, and reattached by Philip Wride, who also built up the topsides leaving the interior surfaces unfinished. No other replacements were made and all the fittings and interior details are original.

CONSTRUCTION

Scale: 1/48 Hull length: 54in

This large model is extremely accurate and is identified by exact correspondence with the measurements of the *Généreux*, which was the only vessel of these dimensions in the Royal Navy. The model may have been built around 1800 when the ship was captured from the French. It is also possible that this model was constructed at a French naval dockyard, as the quality and craftsmanship are extraordinary and not strictly in the English tradition, although it is built to the usual Admiralty Board scale and did turn up in England.

The hull is made with planks of pine about 1in thick, glued together and shaped. The interior is hollowed out to a thickness of

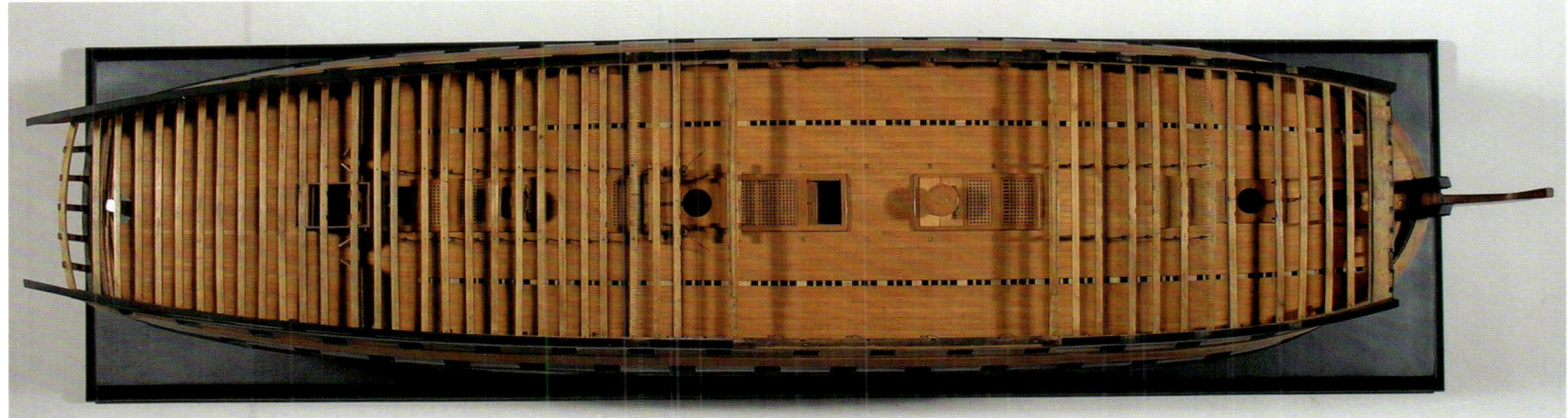

The beam was substantial on this large two-decker, which would have been a relatively commodious vessel.

The upper deck details attest to the high standard of craftsmanship displayed on this model. The beams, bitts, crosspieces, knees and beam shelf are all edged with ebony inlay. The gratings are properly built with battens and ledges, and the metalwork is impressive and includes hinged deck beam pillars, fittings for the elm tree pumps and sheaves for the bitts. The deck planking is virtually seamless.

The wardroom bulkhead consists of partitions and doors whose panels are all inlayed with ebony. The wardroom pantry is seen, surrounding the mizzen mast, with its louvred panels. Removable metal pillars support the quarterdeck beams, and the drumhead of the main jeer capstan can be seen, with the pawls to check its motion, attached to the partners.

¼in at the waterline expanding to 1in at the keel. A false keel is fitted, and the exterior is laid with boxwood strakes perfectly fitted, rabbeted and glued into place with stealers fitted at the bow. A complicated scarph unites the keel and stem. The channel wales are made of boxwood, but the main wales are built with ebony strakes, as are the external planks of the main gun deck.

The care lavished on the interior details is extraordinary, especially because it is so difficult to see. The gun deck is fully planked and is completely and accurately fitted out. The riding bitts are painted red, but there is a finely scored blackened grove near each edge of the crosspieces and standards, and a similar decorative feature appears on each knee, deck beam and carling. An unusual feature of this model is that all of the standard and hanging knees forward of the mainmast rake backward, and those abaft the mainmast rake forward. Additionally, the standard knees extend through the upper deck and terminate at the beam shelf of the quarterdeck. All pillars are finely turned, and shot garlands are fitted to most of the coamings around companionways and gratings.

Reciprocating pumps are fitted before and abaft the mainmast complete with finely wrought brass handles, wooden tubes, cisterns and dales. The jeer capstans have six whelps above and below; with drumheads pierced for twelve bars and trundleheads pierced for eight. The capstan partners are fitted with a pair of brass pawls that engage the fore side of the capstan. Officers' cabins at the stern are fitted with lights and doorways.

The upper deck details are a testament to the abilities of the anonymous builder. There is a thin strip of ebony inlay adorning the fore and aft edges of all the deck beams, knees, the lower edge of the beam shelves, the jeer bitts, topsail sheet bits, and the crosspieces. The spirketting is made of ebony, as are the gun port sills, lintels and frames. The screen bulkhead is panelled with ebony inlay, and there is a pantry and ventilated meat locker amidships. Detailed pumps are fitted with pivoting brake handles linked to brass spears and yoked to wooden casings with brass hoops and discharge ports.

Quarterdeck beams forward of the mainmast and forecastle

THE GÉNÉREUX, A 3RD RATE OF 1785

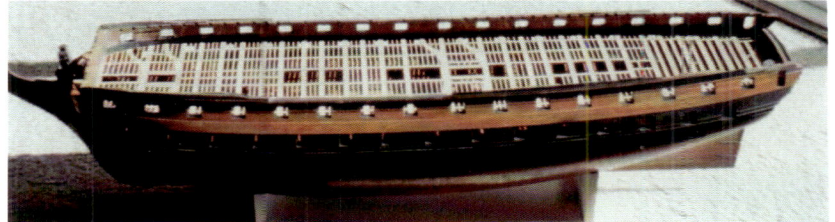

The beam was substantial on this large two-decker, which would have been a relatively commodious vessel.

This photograph was taken while the model was taken apart for cleaning. The upper deck beams remain in place, and chamfers for carlings, ledges and forecastle deck beams can be seen.

The upper deck planking, carlings and ledges have been removed for cleaning. This reveals centreline details including the riding bitts, capstan partners, companionways, scuttles, coamings and shot garlands.

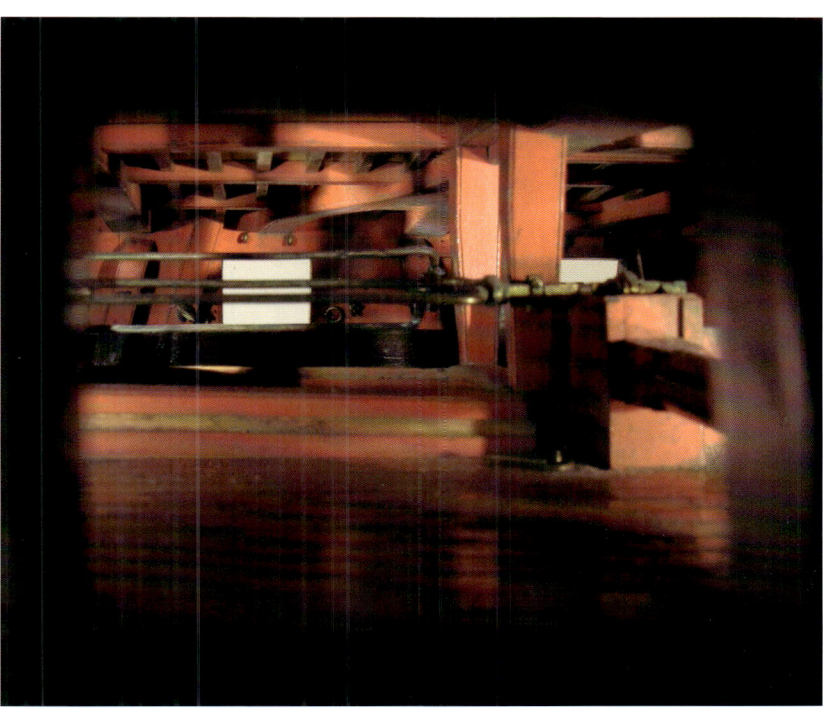

A glimpse through a main gun deck porthole amidships reveals the main chain pumps. On the left are the crank handles, and to the right is the cistern with its wooden dale running to the ship's side. It is hard to believe so much care was taken to construct details that, under normal circumstances, can barely be seen.

beams abaft the foremast are supported by pairs of brass pillars. These pillars are hinged to the beams and slide into metal fittings on the deck so that they could be taken up when the ship was cleared for action. Similar removable pillars supported beams across the waist.

Literature

The following reference includes a photograph and description of this model:
Lavery, Brian and Simon Stephens. *Ship Models: Their Purpose and Development from 1650 to the Present* (London: Philip Wilson, 1995) p. 20.

Exhibitions

Greenwich, National Maritime Museum, *Ship Models from the Great Age of Sail 1600–1850*, 18–20 April 1996.

Historical Perspective

MODELLING A PRIZE

This is one of the large two-decker 3rd rates built during the Napoleonic era. It is also an example of the numerous French-built vessels captured by the British and incorporated into the Royal Navy during that prolonged conflict. She was designed by Sane and launched at Rochefort in 1785 as *Le Ge'ne'reux*.

On 1 August 1798, *Le Généreux*, captained by Le Joille, was with Vice Admiral Brueys' fleet in Aboukir Bay outside of Alexandria, Egypt. They had just transported Napoleon and his army to North Africa and were at anchor in the bay when attacked by Vice Admiral Horatio Nelson. Known as the Battle of the Nile, this was the most complete naval victory achieved by the Royal Navy up to that time. Most of the French fleet were captured or destroyed, and only four French vessels escaped from the bay, among them *Le Généreux*.

Nelson's dispatches to London announcing his victory were sent on board the 50-gun *Leander* under Captain Thomas Thompson. On 18 August, she was sighted by the 80-gun *Le Généreux*. The *Leander* was unable to escape, and a furious action

A glimpse through a main gun deck porthole amidships reveals the main chain pumps. On the left are the crank handles, and to the right is the cistern with its wooden dale running to the ship's side. It is hard to believe so much care was taken to construct details that, under normal circumstances, can barely be seen.

ensued. The uneven engagement continued for over six hours with raking broadsides exchanged on both sides until the *Leander*, totally dismasted and shattered, surrendered. Each ship lost about one-third of her crew, and Edward Berry, the bearer of Nelson's dispatches, was struck in the arm by a skull fragment. This action is a reminder of the purpose and use of the ships of war these models represent, as described in *The Wooden World Dissected*, published in 1708, wherein the following succinct definition is offered, 'It's the great Wooden Horse of Nature, for the Accommodation of all such as want to ride in Post-haste from one World to the other.'[1]

By 20 October, *Le Généreux*, along with her prize, the *Leander*, was at Corfu when a British squadron attacked the island. Once again *Le Généreux* was able to escape, seeking refuge in Ancona, although the *Leander* was later recaptured.

On 7 February 1800, *Le Généreux*, flagship of Rear-Admiral Perrée, left Toulon with a squadron of ships bound for Malta, which was under siege by the British. At dawn on 18 February, she was sighted by the *Alexander* (74) who gave chase. The British ships *Success* (32), *Foudroyant* (80), and *Northumberland* (74) joined the attack, and *Le Généreux*, isolated from her small squadron, struck her colours. Her Commander, Perrée, had been struck in the left eye early in the action, and subsequently his right thigh was shot away, mortally wounding him.

The British lost no time incorporating the *Généreux* into the fleet, and under Captain Manley Dixon, she joined the blockade, rather than the relief, of Malta. The island capitulated on 4 September. The *Généreux* subsequently sailed to Gibraltar, then Port Mahon. In the following year, she assisted in the reinforcement of Porto Ferrajo. She was broken up in 1816.

References

Clowes, Wm Laird, *The Royal Navy a History From the Earliest Times to the Present*, Vol. 4 (London: Sampson, Low, Marston and Company, 1898).

There is no evidence that channels, skids or head rails were ever fitted to this model. Inboard features, however, are shown in extraordinary detail.

Port broadside view of the model showing the foremast channel, which is the only one surviving. The small quarter gallery is a distinguishing feature on many of the early American ships of the line. Gun port lids were never fitted to the model, which is also true of the 74-gun ship model by Joshua Humphreys at the Independence Seaport Museum in Philadelphia.

CHAPTER 17

The *Franklin*, American 74-gun ship c1800

Acquisition

OUR COLLECTION BEGAN WITH English builder's models and eventually expanded to include Dutch and French examples. The most elusive shipwright models created by the naval powers of the Occident are American ones. They virtually do not exist. So it was with great excitement that in 1978 we received photographs of a model that appeared to be American. The photographs were from Merritt Edson Jr, the secretary of the Nautical Research Guild, and they showed an unrigged hull of a large two-decker, in somewhat distressed condition. We were confident it was a builder's model, but the quality and features were neither English, French nor Dutch. It had a blend of French and British features that suggested to us a possible American identity, and so Arnold and I lost no time in arranging to see the model in person. It belonged to Mr Francis Reidy, who showed us the model in autumn 1978. We were intrigued and tried to buy the model from him but could not agree on a price. It remained in the Reidy family until it came up for sale in Philadelphia in 2019, and after forty-one years we were finally able to bring it home and to study it in detail.

Provenance

This model turned up in Philadelphia and was purchased by Mr Reidy, proprietor of McClees Antique Galleries operating in Philadelphia since 1947. It became an heirloom to his son, Frank Reidy. The earlier history of the model is unknown.

Description

CONDITION

The model appears to have spent much of its life without a protective case. It is held together by hundreds of wooden trenails and organic glue, and over time the glue has desiccated and failed in many places. Consequently we found a scattering of loose knees, pillars, beams, carlings, ledges, stairways, etc in the hull, most of which we fished out, and some of these we have reattached. The model has also suffered mechanical damage, especially in the head with loss of the stem, head rails, figurehead, catheads, etc. Frank Reidy fashioned a few missing pieces when he owned the model, but these replacements are obvious. Considering its age, fragility and rarity, the model is actually in a remarkable state of preservation, with almost all of its internal parts and many other key elements preserved. It is supported on a solid wooden plinth, inlaid and veneered, which appears to be original.

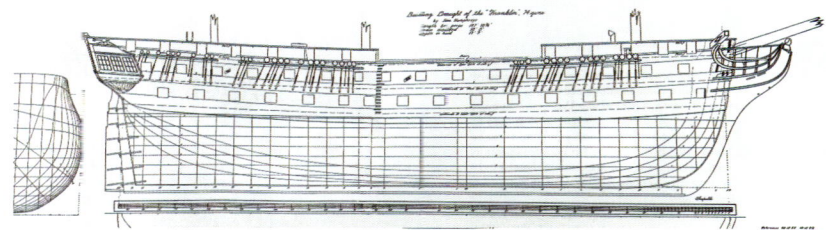

Starboard broadside view showing the *Franklin* mounted on its plinth. Although described as a '74', she is pierced for 80 broadside guns. We believe this model was likely built by Joshua Humphreys c1800 as his initial design for the ship. Below the model is the draft of the ship by his son, Samuel Humphreys, c1814. Despite the time lapse, here is a close correspondence, the chief difference being a rounded bow on the later draft.

The model has no paint or varnish, which it also has in common with the Joshua Humphrey model in Philadelphia. The only metal fittings on the entire model are remnants of the pintles and gudgeons for the rudder.

The external planking is attached to the timbers with wooden trenails, and the model is fitted to a thick but solid wooden plinth, decorated with inlay and veneers.

CONSTRUCTION

Scale: 1/80 (1/8in = 10ft) Hull length: 30in

This model is unique in many respects. Overall, it does not look like an English or French builder's model with respect to the quality of the workmanship, finish, or even the colour. It is neither varnished nor painted, and the carved decoration it does have is very restrained, consisting largely of rope-twist mouldings. It appears somewhat crude compared to its European contemporaries. In fact, it looks just like what it is – an American shipwright's design for a man-of-war built when this country was young. It does not follow any ship modeller's convention with regard to framing or simplification, but rather is built much as the ship it represents. It displays the key structural elements, frames, beams, supports, planks, etc as they would be on the real ship, and is made entirely out of wood with no metal fittings save for pintles and gudgeons (now missing) for the rudder. It is held together with trenail pegs and glue. It is pierced for 80 guns, but the quarter galleries are more appropriate for a frigate, with a single central light, and the stern is likewise of modest proportions with a single closed gallery. The only decoration on the stern consists of simple rope-twist mouldings. There is a fully planked spar deck, but the lower decks on the port side are unplanked, facilitating inspection of the interior.

The outboard ends of the catheads are missing, but the cat-tails are diagonal and overlaid, and while sometimes seen on English ships of this period, it is a feature more commonly found on French vessels. Another sign of French influence is the presence of a fore jeer capstan on the forecastle. The whole is mounted on a thick plinth, inlaid and veneered.

Joshua Humphreys, a colonial naval architect, has been dubbed the 'Father of the American Navy' for his role in designing and building the frigates that catapulted the nascent United States

THE *FRANKLIN*, AMERICAN 74-GUN SHIP c1800

The simple stern with its single closed gallery. The only decorative carvings remaining on the model are simple rope mouldings around the stern lights and quarter galleries.

The square beakhead bulkhead is an eighteenth-century feature that suggests the model was made early on in the design process for the ship, perhaps when originally planned in 1800. When actually built in 1815 it would have had a rounded bow. Most of the original head is unfortunately missing.

Navy upon the world stage. He also played a key role in designing the first American 74-gun ships, and in particular the one built in Philadelphia, the *Franklin*. Furthermore, he was a model builder.

There is another 74-gun ship model, one that was definitively built by Humphreys, preserved at the Independence Seaport Museum in Philadelphia. It is a half-hull model of a 74-gun ship, and on the back of the board to which the model is affixed there is a period inscription that reads: 'Joshua Humphreys Fecit 1777'. On 20 November 1776, the Continental Congress authorised the construction of three 74-gun ships of the line, and this model was made as a design for one of these, but it was never built, as preparations were abandoned when the British captured Philadelphia on 26 September 1777. After the war, the model hung in the mould loft of the Navy Yard in Philadelphia, and was subsequently transferred to Pont Reading, Pennsylvania, home of Humphrey's descendants. It was eventually presented to Independence Hall, Philadelphia, and now resides at the Seaport Museum. It is a handsome and well-preserved model that shares some unusual features with our 74-gun 'Franklin' model. These similarities include: absence of provision for deadeyes in the channels; absence of chainplates; absence of gun port lids; square trenails; and both appear to be made out of the same distinctive unvarnished wood. Key differences are that the Seaport Museum model represents a more conventional 74-gun ship of the late eighteenth century, is twice the size of ours and is a half-model.

Historical Perspective

A MAQUETTE OF AN AMERICAN SHIP OF THE LINE

This view shows how the model has been constructed with internal decking on the starboard side, but with beams and carlings and ledges visible on the port half. The catheads are missing, but the overlaid diagonal tails remain. The flush spar deck is planked and there are no gratings present.

The evidence indicates that this model was built by Joshua Humphreys c1800 as a design for the 74-gun *Franklin*. It is the earliest known builder's model of a man-of-war built for the nascent United States Navy. This conclusion is supported by the following observations:

1. Resemblance to the construction of the autograph model of a 74-gun ship built by Joshua Humphreys in 1777 (see above).

2. Humphreys' apparent use of models in designing the first American frigates
There is evidence in Humphreys' notes and correspondence preserved at the War Department to support his use of models in the early stages of warship design. Models are mentioned in the following documents:

In a letter of 28 June 1794, from Secretary of War Henry Knox to Joshua Humphreys, Knox says:

> Sir, You are appointed the Constructor or Master Builder of a Forty-four Gun Ship, to be built in the port of Philadelphia at the rate of compensation of Two thousand dollars per annum. This compensation is to be considered as commencing on the first of May last, in consideration of your incessant application to the public interests in adjusting the principles of the Ships, drawing of drafts and making of models, &c-

The above letter seems to reference a model that Humphreys had already presented to the War department as there is a letter dated 12 May 1794 that reads:

> Permit me to request the favour of your giving me your opinion on Mr Humphreys model which you have seen, and upon the following points. Mr Humphreys proportions are 147 foot keel- 43 foot beam- 14 foot hold- 7 feet between decks and 7 feet waist, 3 feet dead rising at 2/5. Whether long ships require their extreme breadth as far forward in proportion as shorter vessels and where is the proper place in long ships for the dead flat to be placed? Whether the model has too much or too little raising, and whether it is too sharp or full forward & abaft? Whether the body and after body of the model are proportionable to each other?

Foredeck of the *Franklin* showing the fore jeer capstan and the scuttle for the galley. The fore channel was never fitted with deadeyes or chair plates, and the same is true of the mode of a 74-gun ship built by Joshua Humphreys in 1777.

Another instruction from Knox to Humphreys dated 12 April 1794 reads:

> I request that you would please immediately to prepare the models for the frame of the frigates proposed by you in your letter of this date and also that you would please to prepare an accurate draft and models of the same, the latter to have the frames accurately described.

The following is an extract of a letter written by Humphreys later in his career, clarifying the role Henry Fox had played in designing the first frigates:

> Permit me to observe, on seeing your instructions to Mr Fox, that soon after the commencement of building, I was directed to prepare a Draught and Model for them, the Model was presented to the late Secy at War and is now in your office, in order to make them the most perfect ships the best Shipwrights of this port were called in to give their opinion on the Model, which they did candidly, I was then directed to make such alterations in the formation of the Frigates body as was conformable to the General Ideas …

Models are mentioned in several other similar contexts in correspondence from 1793–94. As construction of the frigates continues over the next several years, models are no longer mentioned, and references are instead only to plans, moulds and drafts. So the 'models' referred to in these documents are neither plans, moulds, or drafts. The question is, are these 'models' three-dimensional wooden artefacts, or do they reference a two-dimensional design? The answer is not clear cut, but the there is a possibility that solid models are being described and were used in the early stages of the design process. Unfortunately, none survive.

3. Resemblance to Humphreys' design for the *Franklin*

The US Navy Department was created under James Adams, and its first secretary was Benjamin Stoddert. He proposed strengthening the fleet with the construction of a host of new vessels, including six 74-gun line-of-battle ships. On 25 February 1799, Congress appropriated funds for the construction of these six two-deckers and six sloops. Design work commenced, and Joshua Humphreys was the principal architect. Subsequently his son, Samuel, and later William Doughty revised and updated the plans. Materials were gathered at six dockyards, but work was halted abruptly with the election of Thomas Jefferson. His administration was not interested in building large and expensive warships, and on 3 March 1801 the Peace Establishment Act placed a hold on their

A series of endoscopic images showing interior details. On the left is the main capstan with barrels on the gun deck and upper deck (with its spindle visible). Removable pillars are present alongside the gun deck barrel. The centre image shows a view looking down into the main hatchway. The Samson posts have steps let into them to serve as ladders, and the floor timbers can be seen. On the right is a view showing the lower part of the fore-topsail sheet bitts.

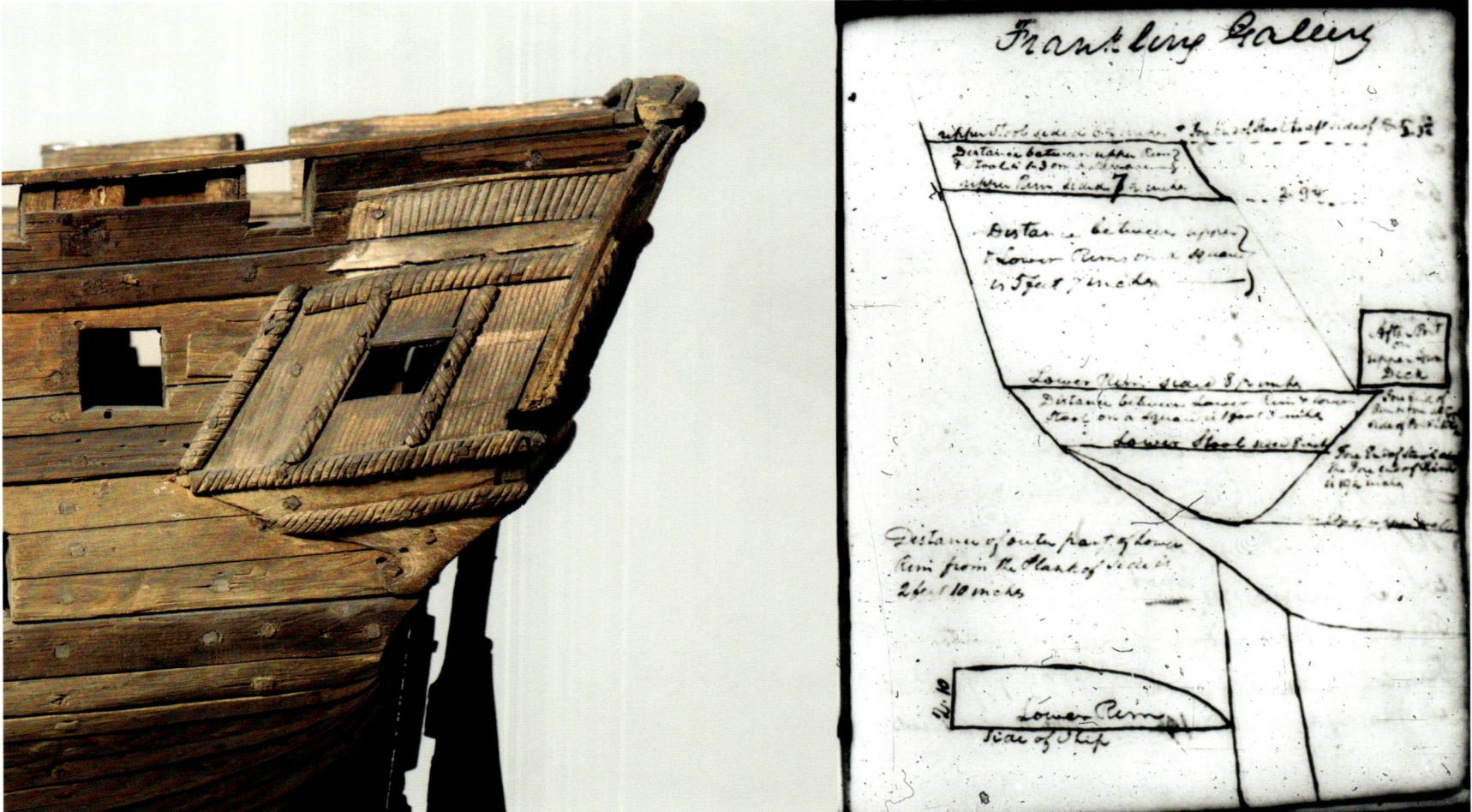

construction. It was at this time that we believe our model was built. The war of 1812 illustrated the value of a strong navy, and on 2 January 1813 Congress passed an Act authorising the construction of 'four ships to rate not less than 74-guns'. When built, these '74s' could carry 86–102 guns. One of them was the *Franklin*.

The *Franklin* was the first naval ship built at the new Philadelphia Navy Yard, and was designed by Joshua and Samuel Humphreys, and built by Samuel. Her keel was laid in 1814 and she was launched on 21 August 1815. A draft of the *Franklin* survives, drawn by Charles Penrose and Samuel Humphreys in 1814, and it is nearly identical to our model. The main and upper gun deck gun port distribution is identical, and both carry eight guns on each side of the 'quarterdeck,' and can accommodate five on each side of the 'fore deck'. As on the model, the draft shows the quarter gallery to be small with one row of lights. Measurements of the model also correspond well with the draft. The draft shows the length between perpendiculars to be 187ft 10¾in, and the beam to be 50ft. On the model, the corresponding length is 29in and

The quarter galleries are simple and would look more appropriate on a frigate, but are just like the one illustrated by Joshua Humphreys in his notebook, shown here courtesy of the Historical Society of Pennsylvania. Unfortunately, this drawing is undated. Some of the hundreds of square-profile trenails that hold the model together can be seen here.

beam is 7¾in, so that the ratio of length to beam on the draft is 3.76, and on the model it is 3.74. The model has a square beakhead bulkhead, and the 1814 draft shows a later round bow. This leads us to speculate that our model was created earlier in the design process, more likely prior to the Act of 1801.

At the Historical Society of Pennsylvania there is a manuscript notebook kept by Joshua Humphreys. It is full of measurements of scantling, beams, yards, masts, etc of ships he designed, and there are a handful of diagrams. By a stroke of good fortune, one of the sketches shows the '*Franklin* Gallery', and it corresponds remarkably well to the model.[2]

An additional piece of evidence exists in the form of a

THE FRANKLIN, AMERICAN 74-GUN SHIP c1800

Among the miscellaneous loose pieces found in the hull are these two with carving on them. The larger one appears to be an intriguing remnant of the figurehead.

painting of the ship as built. Painted by Thomas Thompson in the 1820s or 1830s, it shows the *Franklin* in the Bay of New York. The broadside and stern correspond quite well to the model, save for the open gallery – which appears to be an awkward ex post facto alteration.

4. The model turned up in Philadelphia. Of the four 74s built by the Act of 1813, the *Independence* was built in Boston, the *Washington* was built in Portsmouth, New Hampshire, the *Columbus* was built in Washington, but only the *Franklin* was built in Philadelphia, by Humphreys.

For the reasons above, we believe this model represents a design for the 74-gun *Franklin* and is a unique early builder's model of an American warship most likely made by Joshua Humphreys.

FATE OF THE FRANKLIN

The *Franklin* sailed from Philadelphia on 20 October 1817 to carry the US ambassador to Great Britain, landing at Cowes on 17 November. She served as the flagship of the Mediterranean squadron from February 1818 until April 1820. In 1821 she sailed for South America. In 1822 she was sent round Cape Horn to Chile, where she took command of the Pacific squadron watching over the American whaling fleets. She sailed to New York in 1824 and was placed in ordinary on 14 September. She spent time as a receiving ship in the Boston Navy Yard and was broken up in 1853. She was considered a good sailer, but when fully loaded her lower deck guns were only 4ft from the waterline and unusable in a strong breeze. Fortunately, she never had occasion to employ them.

Detail of a painting by Thomas Thompson showing the *Franklin* in the Bay of New York in the 1820s or 1830s. This painting provides evidence that the ship, as built, featured a simple stern with a small quarter gallery as on the model; although it appears that the lack of an open gallery has been remedied by the ad hoc addition of a balcony supported by external pillars! Photo courtesy of the Metropolitan Museum of Art.

CHAPTER 18

The *Carcass* bomb of 1758

Acquisition

As satisfying as it is to add a new model to our collection, it is always a disappointment when the model has no history. In the case of this model, its location for the last 150 years was a mystery. However, after acquiring it, we were able to discover its identity and a link to Horatio Nelson, a reminder of how models like these are woven into the tapestry of British history. Over the years we have maintained friendly contact with as many marine art dealers as we can on both sides of the Atlantic and made sure that they know our particular interests and have current contact information. In this instance, our efforts were rewarded because we were the first clients contacted once this model surfaced. Faced with the option of buying it directly or seeing it go to auction at one of the major auction houses, there was really no choice. For models as rare as this, simply the opportunity to acquire them renders the cost almost immaterial.

Provenance

This lovely model surprisingly appeared at auction in New Hampshire in 2004. It was purchased at the auction by a dealer and then offered to us. We know nothing of its earlier provenance.

Description

CONDITION

Remarkably, there is no evidence of restoration on this model. It remains in original condition with losses limited to three fenders and three mortar pit support standards.

Left: The mortar pits and corresponding structural accommodations are represented in great detail, along with the shell rooms. Beneath the sheer rail there is a decorative frieze of tendrils carefully painted in gold against a blue ground.

CONSTRUCTION

Scale: 1/48 Hull length: 9in

Several similarly constructed models of midship sections of bomb vessels exist in public collections in England, but none are as detailed or complete as this one. The cutaway style of these models is believed to have been chosen as the best way to reveal the structural modifications necessary to absorb the recoil and support the weight of the mortars that these vessels carried. A 13in mortar, for example, would have weighed about 4 tons.

External features of this model include a band of painted floral decoration beneath the sheer rail, entry port ladders, a solid main wale, and topsides pierced for four gun ports on each side. The hull is fully planked, as are the internal bulwarks, complete with lining, spirketting, waterways and shot racks for the carriage guns.

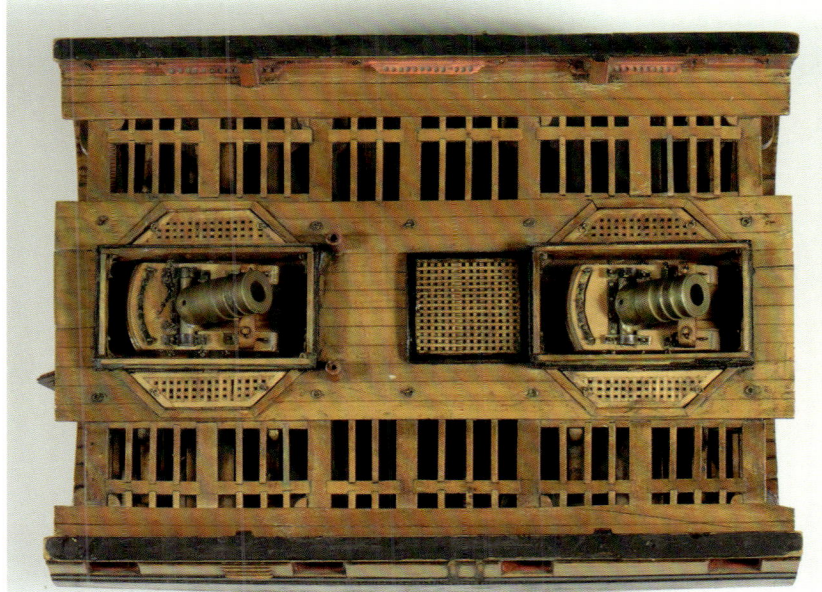

The sea mortars and their beds and fittings are removable, including the cap squares, eyebolts and locking pins for securing the mortars to their beds, the mortar supports, turntable, housings and covers.

This view clearly shows the scarph between the floor and futtock riders, the way the mortar pit planks are ciphered together, and the beams and pillars that support the bed. In the shell room, the racks contain model shells complete with fuse holes and lifting lugs.

The cross-sections of the hull reveal a wealth of constructional details, including the scarph uniting the floor timbers with the second futtocks; the keel, false keel and keelson; the limber board, internal strakes and deck clamps. All of the modifications for supporting the mortars and shell rooms can be studied, including the five floor riders fitted across the keelson for each mortar, scarphed to futtock riders that reinforce the hull to a level just below the mortar pit beams.

Both the fore and aft shell room are fully constructed, each supported on three fore and aft beams resting on the floor riders, with filling timbers in between. Mortices cut into these beams receive tenons at the heels of eighteen square pillars arranged in three longitudinal rows. At the head of each pillar, another tenon can be seen fitted to a mortice cut into another set of fore and aft support beams that form the top of the shell room and reinforce the mortar pit beams. Between these pillars, three tiers of shell stowage racks are fitted, along with the shells. These are modelled accurately with fuse holes, and each shell room has a capacity of forty-five. Each of the heavier shells for the 13in mortar in the aft room are even quipped with a pair of lifting lugs! Entry is through a pair of doors hung on metal strap hinges.

The mortar pits sit upon six athwartship beams supported by hanging knees and the shell room pillars. On top of these beams there is a layer of deck planking carefully rabbeted together, and on top of this there is another layer of plank. The mortar pit is let into this, and each consists of an octagonal well made up of trimmer beams and planks. Each mortar is enclosed in a rectangular housing which is formed of removable sections that are held together by working metal eye hooks. A detachable canopy covers each mortar when stowed. Removable trapezoidal gratings, complete with metal lifting rings and built up with athwartship battens let into cross battens, complete the deck covering.

The mortars and their beds are quite detailed. Each mortar is made of brass and secured to the bed with metal cap squares over the trunnions, complete with eyebolts and locking pins. They are elevated on timber support chocks held in place by metal keep plates and retaining pins, maintaining a 45-degree angle of fire. Each bed rests upon a wooden turntable that is fitted with a central spindle that allows the mortar to rotate into firing position. Abaft the forward pit bulkhead are red casings for the suction pumps.

The *Carcass* was a ship-rigged bomb vessel of the *Infernal* class of 1756. She carried two mortars mounted on the centreline and was designed for shore bombardment. This is an unusually complete model, even including the removable housing canopies not shown in this photograph.

Historical Perspective

THE *CARCASS* BOMB LAUNCHES A HERO'S CAREER

The *Carcass* bomb was built in Rotherhithe by Stanton and launched in 1758. In 1773 she was converted to work as an exploration ship and sent, together with the *Racehorse* bomb, to the frozen reaches of the Arctic. This conversion was not unusual since the hull of a ship that was reinforced to support the weight and stress of large mortars could be easily adapted to withstand the stress of ice floes and pack ice. The voyage to find a north-east passage to the Pacific was endorsed by the Royal Society but ended in failure when the ships encountered an impenetrable wall of ice.

The voyage of the *Carcass* would have been but a small footnote in the annals of naval history were it not for a young midshipman on board making his first significant voyage at sea. The sailor, named Horatio Nelson, had his first brush with fame on this voyage when he attacked a polar bear. According to Southey's biography of Nelson,[1] Horatio and a friend deserted the mid-watch and ventured over the ice in order to hunt a polar bear. The attempt to shoot the bear apparently misfired and the two were saved from the animal by a chasm in the ice and a timely signal gun from the *Carcass* that scared off the bear. Nelson boasted that if he could only have got closer he would have slain the bear with the butt end of his rifle. Nelson attempting to club a ferocious bear with his rifle thus became the first of many heroic deeds immortalised by popular contemporary engravings. The *Carcass* bomb was paid off following the voyage. Nelson next shipped to sea on the *Seahorse*, a 20-gun 6th rate, and headed for the East Indies and the start of one of the most famous careers in the history of the Royal Navy. The *Carcass* was sold out of the service in 1784.

References

Southey, Robert, *The Life of Nelson* (London: Bickers and Sons, 1884).

Details of two drawings from a French manuscript c1690 by le Chevalier de Fabregues, who was an officer of artillery at Brest, France. Bomb vessels were introduced by the French at the end of the seventeenth century, and the two examples shown on the left are anchored in place and firing at an enemy fortress. The ship-rigged vessel to the left has shifted the foremast shrouds aft so as not to interfere with the projectiles. The drawing to the right shows this manoeuvre in more detail, as well as the use of a chain forestay in place of a more flammable rope. The fore yard has been lowered to clear the arc of fire. The mouth of a mortar can be seen just above the foredeck bulwarks.

CHAPTER 18

The *Aetna* bomb of 1776

Acquisition

THIS INTERESTING MODEL WAS presumably intended to display the special constructional details required to support the use of the two mortars. Among our dockyard models, this is one of the most prosaic in that there are really no decorative details, it is all about the business of the mortars, yet there is beauty in the way the modeller has included even the smallest constructional detail, and the use of ivory attests to the pride he took with his creation. When this model appeared for sale in 1983, it was erroneously catalogued as a midship section of a frigate, and was furthermore described as dating from the early twentieth century. Inspection of the model prior to the sale convinced us that it was a period mid-eighteenth-century model and revealed it to be a ship-rigged bomb vessel. When Henry questioned Sotheby's cataloguer about it, he agreed that it appeared to him to be older than early twentieth century, but that when they had asked the authorities at the NMM for a determination they were politely refused on the grounds that the museum might try to buy this model and wished to avoid any conflict of interest. Sotheby's decided to be conservative, and the consignor was agreeable. As it happens, the NMM did not succeed in buying the model.

Provenance

This model appeared for sale at Sotheby's, London, in 1983.

The *Aetna* bomb was launched on 20 June 1776, and was expected to help subdue the rebellious North American colonists. The model is unique among contemporary examples in showing so much of the hull and in being asymmetrically constructed to better show constructional details. The starboard side is fully planked and fitted with channels, deadeyes, chainplates, skids, etc. Below decks, cabins and storerooms extend to the midline only.

Description

CONDITION

When this model was purchased in 1983, there were no miniature mortars mounted in the pits. This, no doubt, contributed to it being misidentified as a midship section of a frigate, which worked to our advantage when the gavel fell. The mortars in place now are modern, but there is no other reconstruction and the model is remarkably well preserved.

CONSTRUCTION

Scale: 1/48 Hull length: 16¾in

Of the half dozen existing models of bomb vessels, this is the only one to show so much of the hull. The model begins just fore of the belfry and extends to aft of the quarterdeck bulkhead, spanning six gun ports (from station 19 aft to station 1 forward). It is also unique in being longitudinally asymmetric, with the starboard side planked inside and out and showing cabin bulkheads and deck supports. The port side is left in frame with only deck clamps, spirketting, outer deck strakes and waterway fitted.

Construction of the keel is represented in detail with the upper and lower false keel, true keel and hog in addition to the keelson. On the port side, the upper and lower cills framing the gun ports are finely fitted with angled joints that allow the upper cills to support the weight of the filling top timbers. A canted timber helps frame the foremost port. The boxwood planking on the starboard side is done superbly. The plank ends are butted together with three strakes between them where they come on the same line. All the seams are uniform and darkened, so that they appear scored. The sheer rails are finely moulded and the aftmost gun ports are hung with lids. Chain plates are bolted to the hull and connected to the deadeye strops by links of chain.

Deck details include pumps, built-up gratings with

The forward mortar is shown stowed with its canopy and side covers in place, whereas the aft mortar is elevated for firing. There is a Spartan absence of decoration, with only a green frieze strake aft of the main channel.

delicately cambered ledges set into frames with half-lapped corners, and a double main capstan. This last fitting is significant, as it compensates for the lack of a jeer capstan, displaced by the twin mortar pits. The balusters beneath the breast rail of the quarterdeck and the casing heads and discharge ports of the pumps are made of ivory.

There is an impressive amount of internal detail, painstakingly fitted and beautifully finished. The limber boards athwart the keelson can be seen to consist of short planks butted together and set into a longitudinal groove in the limberstrake to form the roof of the limber passage. The limberstrake is pinned to each frame with a trenail, and the internal planking continues on the starboard side with the adjacent footwaling, which in turn butts against the ceiling strakes. Two additional rows of thickstuff and footwaling indicate the positions of the scarphs of the floors and futtocks. Above the deck, the waterway, spirketting, lining strakes and sheer rail are all carefully fitted and glued in place.

Forward of the shell rooms, there is a platform with finely turned pillars that support upper deck beams, and there are storerooms below. Aft of the shell rooms there is an after hold and then the after platform with a central lobby and two officers' cabins.

The shell rooms are built in the standard manner, but there are four tiers of shell racks in the forward room, whereas there are three in the aft room. Since the *Aetna* carried a 10in mortar in the forward pit and a 13in one aft, the larger shells left room for only three tiers in the aft room.

~ Literature ~

The following reference includes a photograph and description of this model:

Gardiner, Robert, *Warships of the Napoleonic Era* (Barnsley, Yorks, Seaforth Publishing, 2011), p. 93.

Historical Perspective

FIRE AND BRIMSTONE, WOOD AND CANVAS

This model represents the *Aetna* class of bomb vessels ordered built in 1776 to help deal with the rebellious American colonists. Two bomb vessels were built originally, and the order was repeated in 1778 when the French entered the war. All four were ship-rigged and carried two mortars, one of 10in bore and one of 13in. The mainmast was abaft both mortar beds, and they all closely resembled this model. British bomb vessels all bore wonderfully evocative names, often borrowed from famous volcanoes, and these four were *Aetna*, *Vesuvius*, *Terror* and *Thunder*. They were built with remarkable expediency, and *Aetna* was launched on 20 June 1776, having been laid down in March of that year. *Thunder* foundered in 1781, and *Aetna* was broken up in 1784, but *Vesuvius* and *Terror* survived to see action against France in the Napoleonic Wars and were both sold out of the Navy in 1812.

References

Ware, Chris, *The Bomb Vessel* (Annapolis, Maryland: Naval Institute Press, 1994).

The framing of the port side is unplanked, showing fore and aft ribands, and gun port sills and lintels. Note how one of the top timbers is canted to form the side of the foremost gun port. The stanchions of the quarterdeck breast rail, and the casing heads and discharge ports of the elm tree pumps, are made of ivory.

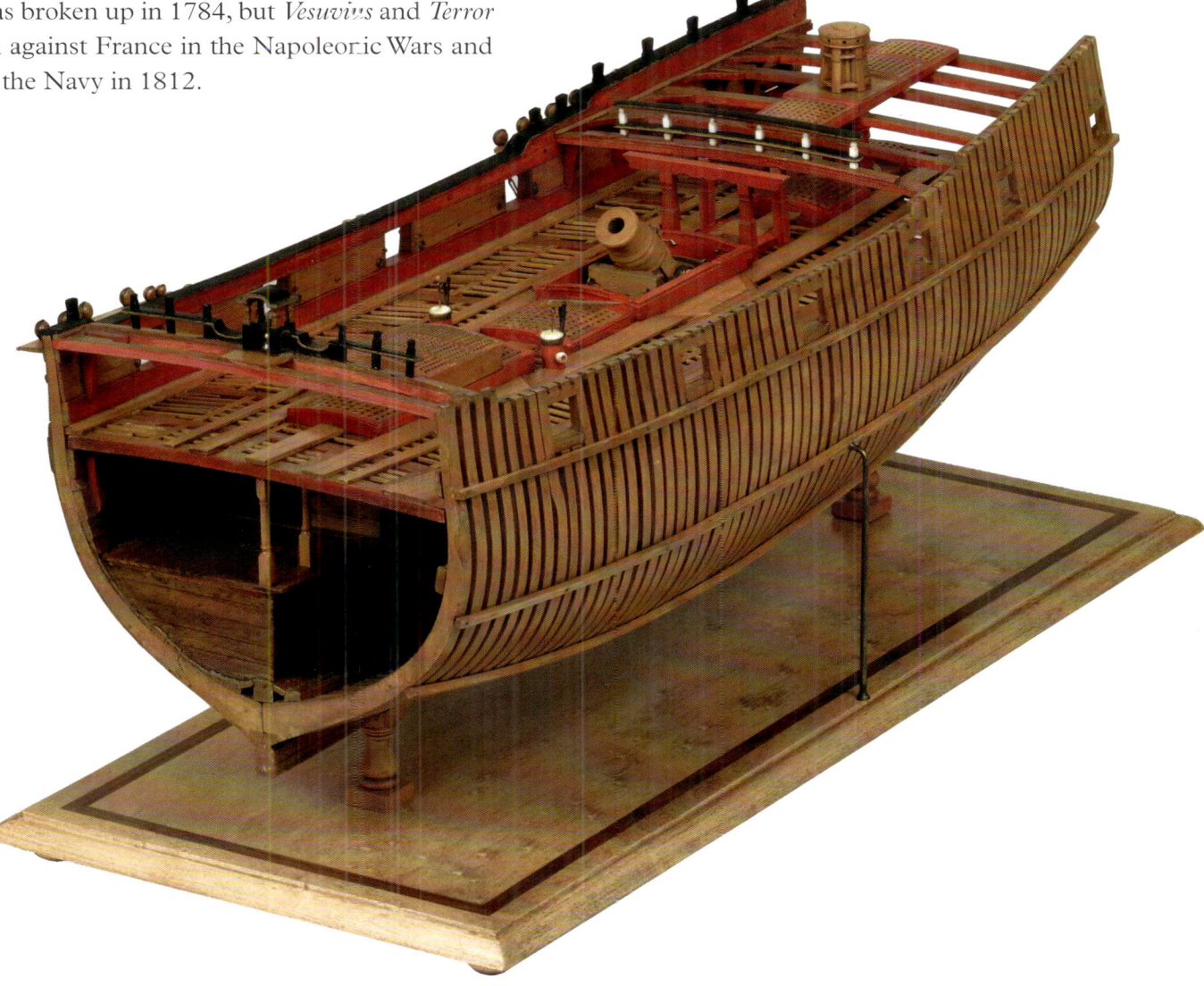

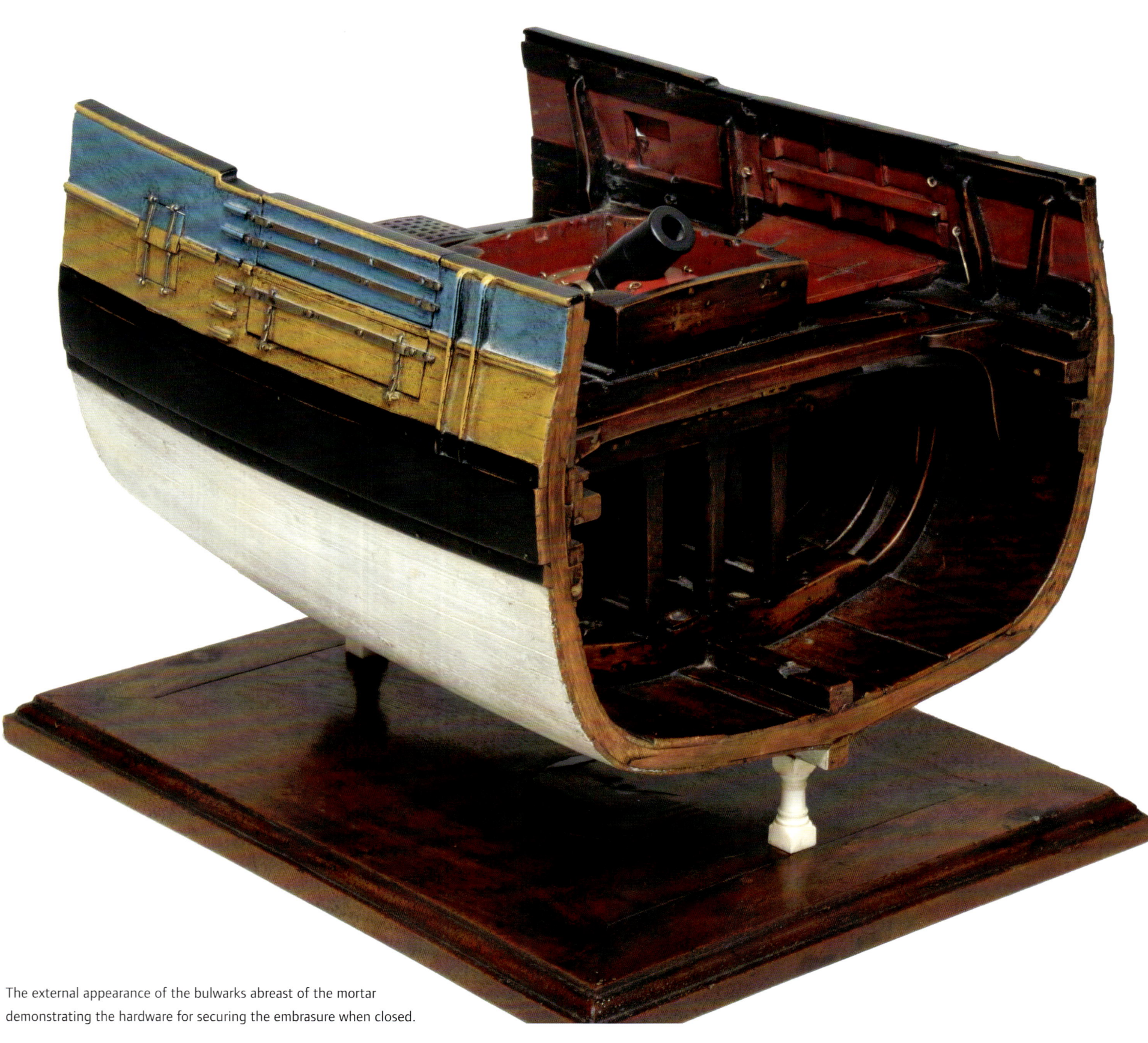

The external appearance of the bulwarks abreast of the mortar demonstrating the hardware for securing the embrasure when closed.

CHAPTER 20

The *Sulphur* bomb of 1797

Acquisition

A SURPRISING NUMBER OF dockyard models were made of the midship sections of bomb vessels during the second half of the eighteenth century. This example came to our attention at a Christie's auction held in Paris in 2010. It was presented to us by David Thomson after the sale. As an observant reader will note, we now have three bomb vessel models. Despite their similar subject matter, they are each very distinctive and comparisons among them are interesting and instructive.

Provenance

This model was part of the collection of Charles de Langlade, a native of Marseille who became a Parisian antiques dealer in the 1980s. We acquired this model when his collection was sold by Christie's, Paris, in June 2010.

Description

CONDITION

The model is complete. The paint has been refreshed in places at some time in the past.

CONSTRUCTION

Scale: 1/48 Length: 8in

This is a contemporary model of the midship section of a bomb vessel. It shows a single mortar pit and the underlying timber supports and shell room. The model represents the aft mortar, situated just before the mainmast. The external bulwarks are fitted with entry steps and skids. The mortar is in place in its bed with a removable hatch cover. An unusual feature is the top- and bottom-hinged gun port just aft of the mortar and, most interesting, the hinged, folding bulwarks. The bulwarks in the way of the bomb beds were made to collapse in order to provide an embrasure for firing the mortar at low angles. This innovative feature first appeared on this class of bomb vessel and may be the key feature this model was made to illustrate. By 1808, however, it was decided to eliminate the embrasures since in practice there was no need to fire the mortars at such low angles.

The 10 in mortar, its bed and supporting pillars are seen in this view, which also shows the midship section.

Literature

The following reference includes a photograph and description of this model:

Gardiner, Robert, *Warships of the Napoleonic Era* (Barnsley, Yorks, Seaforth Publishing, 2011), p. 98.

The port and starboard embrasures work, as can be seen in this photograph where the hinged bulwarks have been lowered to permit firing the mortar at a low angle.

Historical Perspective

COPENHAGEN BOMBARDED

We have identified this model as the midships section of the *Sulphur* bomb vessel of 1797. It matches exactly the sheer and profile draft of the *Sulphur* as converted from a merchant sloop bought into the service in 1797. Although she also resembles her sister ships, *Explosion*, *Strombolo*, *Hecla*, *Tartaru*, and *Volcano*, which were all sloops converted to bomb vessels in 1797, minor differences make any of these other possibilities less likely. She was manned by a crew

of sixty-seven, and in addition to a pair of 10in mortars, she was equipped with 4 6pdr cannon and 6 18pdr carronades. Small details, for which there is precise agreement on both the model and draft, include the presence of two bottom-opening hinges on the embrasure port (whereas three are indicated on drafts of her sister ships) and the presence of skids just forward of this port. The *Sulphur* was hulked in 1805 and sold out of the service in 1816.

While the *Aetna* (see Chapter 17) was one of the bomb vessels built to help quell the American Revolution, the *Sulphur* and her sister ships were a response to the French Revolution. She saw service against the Spanish in Aix Roads in 1799 and participated in the Battle of Copenhagen in 1801 under Vice Admiral Horatio Nelson. It was on the occasion of the latter battle that Nelson famously refused to retreat as ordered by a signal from Admiral Hyde Parker's flagship. Standing next to Captain Thomas Foley, Nelson held his telescope to his blind eye and stated, 'You know, Foley, I only have one eye — I have a right to be blind sometimes,' and then, 'I really do not see the signal.'

The fine detail and fittings of the mortar bed and pit can be seen, along with the inboard modifications for the embrasures.

The high sheer to the stern of this ten-oared barge adds considerable grace to the lines and is characteristic of early eighteenth-century barges. The colour scheme consists of varnished boxwood, gilded mouldings and carvings, and red interior surfaces, and matches that of warship models of the same period (see Chapter 6).

CHAPTER 21
A Queen Anne royal barge

Acquisition

Several models in our collection, including this little one, were unknown to us until they appeared at auction. For many years there was relatively little interest in these exquisite little examples of the shipwright's craft, and over time we very happily acquired a small collection of dockyard models of ship's barges. This lovely barge model appeared at auction at Sotheby's on 28 October 1986. Arnold was in London at the time and was able to buy it and bring it back to New York.

Provenance

When this model was purchased at auction there was, regrettably, no history of prior ownership.

Description

CONDITION

It is surprising that such a delicate little barge model has survived in such remarkable untouched condition, including three of its original oars. It must have been kept in a storage box or display case. Only three sweeps, the rudder, and the sternsheet step were missing, and Philip Wride replaced these. The walnut-veneered plinth and metal cradles are modern.

CONSTRUCTION

Scale: 1/24 Hull length: 19½in

This is a contemporary Admiralty Board model of a ten-oared barge, built in the time of Queen Anne. The model, made of boxwood, is of open-frame construction and represents a carvel-built barge. There are ten thwarts to seat rowers for ten single-banked oars. The sternsheets are fairly typical, equipped with a thwartships bench backed against the backboard and two longitudinal benches extending forward. Lockers are fitted underneath the benches accessed by circular openings in their forward ends. A gangboard extends forward along the centreline, but is composed of separate segments, each fitted between two consecutive thwarts and designed to slide to one side to provide accommodation for the rowers. Consecutive gangboard segments slide alternately to port or starboard, and the craftsman has gone to the trouble of decorating the non-sliding edges of the thwarts

The commodious stern sheets can accommodate up to seven passengers comfortably. Lockers are provided under the benches. Gilt mouldings outline panels on the backboard and inner gunwales that may have contained decorative paintings or the actual barge. Holes drilled into the tops of the gunwales are provisions for a removable awning.

The voluptuous female bust on the transom most likely represents Queen Anne herself. She is surrounded by a wreath with a pair of Tudor roses flanking the socle. Note the lunate arch of the transom and the lateral cut-outs to accommodate the gunwales with their decorated quarter pieces.

secured by two trenails. There are twenty-six frames, and strikingly, ebony futtocks are fitted afore of frames four to eleven and abaft of frames twelve to twenty-two, with the transition between frames eleven and twelve marked by a markedly reduced room and space. The sternmost frame is canted at an angle parallel to the sternpost and straddles the deadwood. The rudder is hung on the sternpost by two gudgeons and pintles, with the upper pintle facing downward and the lower facing upward. There are coxswain seats to both starboard and port. The deadwood at the bow extends to frame four, the furthest reach forward of the keelson, and the keelson ends beneath the sternmost thwart at frame nineteen. Two

Despite the modest size of this little vessel, the decoration is consistent with that found on full-size ships of the period including carved and gilded acanthus leaf decoration along the sheer strake, gilded hancing pieces and quarter pieces, and a finely carved and gilded transom. One plank is fitted below the gunwale, and the garboard strake and first plank are rabbeted to the keel and pinned to several of the frames.

and gangboards with a finely made stepped moulding. He has even chamfered the edges so that the moulded edge is continuous when the boards are aligned along the centreline. The thwarts are grooved to accept the sliding gangboards, and the boards have wooden fittings underneath to grasp the undersurface of the thwarts. Panels framed with gilded mouldings decorate the insides of the gunwales and the forward face of the backboard. The fore part of the sheer strake is painted red and trimmed with gilded moulding, while the aft part has carved frieze decoration in the form of gilded floral vines with the transition marked by a hancing piece carved with a sprig of roses. The transom is decorated outboard with a robustly carved portrait of Queen Anne flanked by laurel leaves sporting Tudor roses. The transom is bordered by quarter pieces decorated with stylised lion masks trailing leafy twigs that end with Tudor roses. A finely carved crowned cipher of Queen Anne decorates the transom's forward surface.

The stempost and keel appear to be fashioned from one piece, and it is fitted to the sternpost with a mortise and tenon joint

A QUEEN ANNE ROYAL BARGE

The very finely carved and gilded crowned cipher of Queen Anne, seen here with the rudder unshipped, helps to date this model to her reign. The seats in the coxswain's cubby permit steering with either hand.

strips of footwaling are fitted along the floor beneath the thwarts, as well as notched planks to take the kick boards to support the rowers' feet. A single unscored plank represents the sternsheet footwaling. The garboard strake and first plank are present, fashioned from one piece pinned to the frames, and there is one plank fitted below the gunwale. The deadwood at the bow curves up to join the stemson, and a strip of brass pinned to the leading edge protects the stempost. Consistent with the early eighteenth-century date of this model, there is a carved and gilded bracket on either side at the bow in the form of a male torso trailing into floral decoration with a rose at its termination at the stempost. The breast hook has an unusual curved slot in the midline and incorporates an aperture for the jackstaff.

Literature

The following reference includes a photograph and description of this model:
Kriegstein, Arnold and Henry, 'The Kriegstein Collection of British Navy Board Ship Models', *Nautical Research Journal*, Vol. 38, No. 4, December 1993, pp. 221–22 and plate 9.

Historical Perspective

AN ADMIRAL'S BARGE CONVEYS SIR CLOUDESLEY SHOVELL FROM ONE TRAGIC FATE TO ANOTHER

On 22 October 1707, at around 8pm, on a cloudy, wet, but otherwise unremarkable night, the 2nd-rate English warship the *Association* struck rocks off the Isles of Scilly and sank. In quick succession the *Eagle*, *Romney* and *Firebrand* shared the fate of the *Association*, with the loss of 2,000 sailors.

The illustrious Sir Cloudesley Shovell, admiral on board the *Association*, was among those who died on that day, but he did not go down with the ship. Sir Cloudesley died on shore in a little sandy cove called Porth Hellick Bay, on St Mary's island some 7 miles from the wreck site. There is a well-known story about how the admiral met his untimely end.[1] When the body was recovered on the beach, it was found to be missing several rings, including a very valuable emerald one that the admiral always wore. The whereabouts of the ring remained a mystery for many years, until an old woman, a native of St Mary's, confessed on her deathbed not only to taking the ring, but also to dispatching Sir Cloudesley, whom she found alive on the beach. The stolen ring was recovered by a clergyman, who subsequently presented it to Lady Shovell. She set it in a diamond pendant. On her death, the jewel was bequeathed to Lord Dursley, who had been a very close friend of Sir Cloudesley's, and as captain of the *St George* had narrowly avoided sharing Sir Cloudesley's fate on that night of 22 October. Lord Dursley had originally presented the emerald ring to Sir Cloudesley as a token of friendship. Dursley later became Lord Berkeley, and the ring remains in the Berkeley family to this day.

Sir Cloudesley's body was disinterred from its shallow grave on the beach and was brought to London, where it now lies beneath a memorial in Westminster Abbey. The bones of the *Association* lie scattered on the ocean bed. Several accounts over the years have suggested that the illustrious admiral was as heroic in death as he had been in life, such as the conjecture that he bravely swam to Porth Hellick Bay with the help of floating debris, only to succumb to exhaustion upon reaching the beach.[2] But there was sufficient forensic evidence recovered in Porth Hellick Bay in the days following the tragedy to suggest that the admiral's final desperate act may have been far more prosaic. It apparently also involved the ship's barge.

Recovered on the beach along with the admiral were the bodies of: Captain Loades, the commander of the *Association*; Captain Whitacre, the captain of the *Association*; Henry Trelawney, a son of the Bishop of Winchester; several other sons of nobility; and Sir James and John Narborough, who were Sir Cloudsley's stepsons. Also discovered nearby was a chest belonging to Sir Cloudsley, as well as Cloudsley's pet greyhound, and an item that most concerns the present narrative, the stern of Sir Cloudsley's barge.

One can well imagine those last desperate minutes in the life of the *Association*. Three guns were fired at the time the doomed ship struck the rock, and amidst the panic and confusion Sir Cloudsley must have lost no time arranging for his chest and favourite dog to be placed in his barge (though the dog may have acted on his own). He also presumably assembled the coterie of grandees later found in Porth Hellick Bay, before calling for the barge to be lowered away. The barge travelled the 7 miles to the island of St Mary's, where it may have been beaten to pieces by the rocks and surf, thereby scattering the luckless occupants on the beach. Or they may have landed safely, only to be greeted by a welcoming party from St Mary's who may have dispatched the survivors and relieved their bodies of any valuables they had carried with them. In any event, it is clear that in this instance, the captain did not go down with the ship. All of the remaining crewmen, approximately 900, did perish with the ship, and bits and pieces of the wreck continued to wash on to the island for many days afterward.

The tale of how four Royal Naval ships came to sink on a nasty but stormless night, is an infamous one. Sir Cloudesley Shovell was the admiral and commander-in-chief of the Mediterranean fleet in 1707. As the campaign season drew to a close with the approach of winter, Sir Cloudesley prepared to return home with a fleet consisting of fifteen ships of the line, four

This is a rare example of the transom or sternboard of an admiral's barge c1690, bearing the cipher of King William and finely carved reclining figures, trophies of arms and dolphins. The shape, with provision for the gunwale on either side, matches the shape of the sternboard on the model.

fireships, one sloop and one yacht. On the homeward voyage, the weather became thick and stormy, and for several days it was impossible to take a sighting to determine position. Finally, on 21 October, the weather moderated, and on the 22nd, Sir Cloudesley convened a meeting of all the sailing masters to determine the position of the fleet. The Master of the *Lenox* firmly believed that the fleet was off the coast of England, three hours' sail from the Isles of Scilly. The rest believed that they were off Ushant, near the coast of France. Sir Cloudesley adopted the majority view, and plotted his course accordingly. Unfortunately, he was mistaken.

Navigation was severely limited by an inability to determine longitude at sea, a determination that requires accurate time keeping. One consequence of the tragic wreck was that Parliament established a prize to be awarded to the individual who could design a device that would tell longitude accurately at sea. The reward was calibrated to the accuracy of the device and ranged from £10,000 for a method accurate to within one degree to £20,000 for accuracy within one half a degree. John Harrison finally met and exceeded the required accuracy in 1761 using an ingenious spring-driven clock, but was not awarded the £20,000 prize until 1773, and then only when the King intervened on his behalf.

Sir Cloudesley Shovell in the *Association* ran on to the Scilly rocks due to a mistake in reckoning longitude. This painting shows the moment when the *Association* struck the Gilstone ledge. She can be seen firing distress signals, but in four minutes, she sank with loss of all her men. Sir Cloudesley and a select company of officers, however, put off in the admiral's barge, seen suspended in the ship's waist. Sir Cloudesley made it to shore 7 miles away, only to be murdered for his ring.

CHAPTER 22

An admiral's barge c1710

~ Acquisition ~

ONE CAN ONLY SPECULATE on what motivated the English craftsman who made this little model 300 years ago to people it with a full complement of bargemen. Perhaps it was a convenient method of mounting the oars, but this would hardly explain the admiral in the sternsheets or the coxswain manning the tiller. The Navy Board modellers seem to have enjoyed peopling their barge models with wooden crewmen since there are several other examples of barges fitted out this way (see Chapter 22). In 1993, the Edward James Foundation decided to sell this model, along with another small barge model and a model of a Queen Anne two-decker, at auction at Christie's. We were able to purchase this barge at the auction on 7 October 1993. Interestingly, many years earlier Arnold had acquired a pastel portrait of Admiral Sir John Jennings, the original owner of the model, portraying him in his later years. There is actually a striking resemblance between this portrait and the carved, younger, admiral in the sternsheets of the model. We suppose the resemblance is coincidental, but nonetheless still wonder if the distinguished passenger is meant to be Sir John himself.

~ Provenance ~

The original owner of this interesting model was Admiral Sir John Jennings, Lord of the Admiralty in 1714–27. The model passed through his estate to the Edward James Foundation. The foundation placed the model on loan to the NMM from 1983 to 1993.

Above: Years before acquiring the barge model, we bought the pastel portrait of Admiral Sir John Jennings, illustrated here showing him in middle age. Coincidentally, he was the original owner of the barge model, and if one imagines him in his younger years bewigged in black curls, he would bear a striking resemblance to the admiral in the barge.

Left: This single-banked oared barge with its twelve bargemen, coxswain manning the tiller, and admiral in the sternsheets, is one of a handful of such model barges that were animated with little wooden sailors.

Description

CONDITION

This barge model has survived in remarkable original condition. It is one of the very few to be fitted with, and to have retained, its original rowers, coxswain and passenger. The model has an excellent original patinated gilded and polychrome finish, including the finely painted costumes of the carved wooden figures. In addition, the original Union Jack, painted on heavy stock paper, still flies at the jackstaff crowned with a gilded acorn finial. All twelve oars are original, but only ten of the original oarsmen have survived. The two missing figures were expertly replaced after we acquired the model.

CONSTRUCTION

Scale: 1/20 Hull length: 22in

This is a contemporary Admiralty Board model built at the Royal Dockyard in the time of Queen Anne. The open-frame construction is of boxwood, and there are both polychrome and gilded decorations including a carved floral motif along the bulwarks.

The dress of the figures, stylish in their day, is a reminder that this little barge is 300 years old. It is in remarkable original condition and quite colourful with red lacquer interior surfaces and panelling highlighted in gold.

The admiral (or captain) is seated primly in the sternsheets, bewigged, wrapped in his boat cloak to protect against the cold, and sporting a tricorn hat. A dutiful coxswain dressed in appropriate livery mans the tiller. The rudderhead bears a crowned effigy of Queen Anne.

Most remarkably, there are realistically painted carved wooden figures of twelve bargemen complete with red-lacquered oars decorated with painted dolphins on the blades; a coxswain in livery steering the barge; and an admiral or captain in the sternsheets wearing a periwig, a three-cornered hat, cravat and red boat cloak.

This is an example of a barge carried on board a ship for use by a flag officer. The barge is carvel-built and has twelve thwarts for single-banked oars with one rower to each thwart seated on the side opposite his oar. This provides a shorter arc of motion but greater power per stroke than for double-banked oars. The gangboards are made so that they can slide to one side, and the thwarts are grooved along the appropriate starboard or port side to accommodate this motion. The centre thwart is wider than the others and drilled to support a mast. The keel is rabbeted to receive planking and a keelson has been fitted only between the second and eleventh frame. The stem and sternpost are joined to the keel with plain scarphs. The sternpost has a considerable rake and is backed by a false-post and a deadwood knee, while the stempost is backed with a stemson. Sixteen frames are fitted, and the maker has numbered them all in pencil on the portside; the foremost five are numbered on the foreside of the frame, and on the overlying strake, the sternmost six are numbered on the edge midway up the futtock. Some of the numbers are upright; others are sideways and all are clear and pristine.

The stern badge is carved with two cherubs holding aloft a crown. The quarter figures are male caryatids. Two lapstrake planks are pinned to the hull below the gunwale.

The sternmost frame is of the same scantling as the others but forms the fashion piece and is raked and fastened to the sternpost. The first and last two frames fore and aft are made of two pieces combining floor and futtocks, one on each side, of single thickness joined at the centreline, but the mid-section frames consist of floor and futtocks overlapped up to the stringer and are therefore of double scantling. The midsection frames are curiously notched outboard near the keel as though to accept an extra-thickness garboard strake. Two planks are fitted below the gunwales and are neatly secured to each futtock by pairs of brass pins. The gunwales are decorated outboard with a frieze of sinuously carved gilded acanthus leaves and inboard with a series of gilt-framed panels painted with decorations of serpentine vines highlighted in gold on a red ground. On either side of the stem at the bow, there is a carving of a standing figure wearing a cap and apparently holding a dolphin suspended by its tail. Given that the cap appears identical with that worn by the oarsmen, the carving might allegorically imply that the sailors are masters of the sea. There are also carved figurative terms either side of the bow, curving forward at their tail ends to terminate at the stem. In the triangle thus formed just aft of the stem, are a cluster of three carved garter stars. The significance of this latter detail is unclear to us, other than to symbolise the order of the garter of which the sovereign was the primary member. The gunwales are fitted with twelve rowlocks of standard type.

The rudder is hung on the sternpost by two gudgeons and pintles. The upper pintle faces downward and is attached to the rudder, while the lower one is longer, faces upward and is attached to the sternpost. When shipping the rudder, the lower gudgeon is first placed over the lower pintle and lowered before the upper pintle is guided through the upper gudgeon. The rudder is decorated along its upper sides with carved and gilded gadrooning. The tiller is a gilded twisted metal bar with an upturn at the end terminating in a ball. A smartly dressed coxswain in a brown coat with a red cap and white cravat is seated on the gunwale holding the tiller. The tiller head terminates in a crowned gilded female head, presumably an effigy of Queen Anne. The transom is decorated outboard with two carved cherubs holding aloft a golden crown. In keeping with the elaborate decorative scheme characteristic of the early eighteenth century, there are carved figurative terms either side of the transom.

The footwaling is made from a single piece and steps up to the level of the sternsheet, which is also of one-piece construction. The sternsheet benches are fitted with large circular ports in their fore ends for storage. The backboard is decorated with a gilt mould-framed panel painted with a trophy of arms. A bewigged admiral looking rather smug in a red boat cloak and tricorn hat is seated on the sternsheet bench.

The oars vary in length according to position, but all have blades decorated with painted dolphins, shanks that are square in section, round looms and smaller-diameter handles that pass through holes drilled in the hands of the wooden oarsmen. The rowers are all dressed identically with red caps, white tunics, red trousers, grey stockings and black shoes.

Literature

The following reference includes photographs and a description of this model:

Kriegstein, Arnold and Henry, 'The Kriegstein Collection of British Navy Board Ship Models', *Nautical Research Journal*, Vol. 38, No. 4, December 1993, p. 221, plate 8.

~ Exhibitions ~

Greenwich, National Maritime Museum, 1983–93.

~ Historical Perspective ~

THE INVASION OF ENGLAND IS AIDED BY A SHIP MODEL COLLECTION

Barges, such as the one represented by this model, were in frequent use on the Thames and its tributaries throughout the seventeenth and eighteenth centuries, ferrying officers and Navy Board officials between ship and shore. There was one famous occasion, however, when a barge like this was used to transport a collection of Admiralty Board ship models up river in order to keep them safely out of foreign hands. In 1667, while negotiations to end the Second Anglo-Dutch War were dragging on in Breda, the Dutch admiral Michiel de Ruyter launched an audacious attack to sink or capture the most powerful ships of the Royal Navy, which, in view of the ongoing peace process, were laid up in ordinary in the River Medway.

On 7 June, a squadron led by van Gendt joined seventy sail of ships commanded by de Ruyter near the mouth of the Thames. On the 10th they captured the fort at Sheerness and burned the storerooms. The British were panicked, and King Charles ordered Monck, the Duke of Albemarle, to Chatham to strengthen the defences. The best defence against a raid on the higher reaches of the river, where the capital ships were moored, was the great chain stretched between the banks at Gillingham and Hoo Salt Marches. This massive chain was supported by four wooden floats, and Monck sank ships to further protect the chain and the entrance to the Medway. But at high tide on 12 June and with a strong easterly breeze, the Dutch fireship *Pro Patria* breached the chain and the Dutch entered the Medway. In passing, they burned three previously captured Dutch ships that were protecting the chain. Six men-of-war and five fireships continued on the favourable wind and tide all day on the 13th, but when they reached Upnore Castle they came under fire from both banks. No ships were lost, but many Dutch seamen were killed. The Dutch, however, burnt the *Royal Oak* and severely damaged the *Loyal London* and the *Royal James* before retreating down river. Lediard records in his *Naval History* that one Captain Douglas, who was charged with protecting the *Royal Oak*, was heard to say, 'It should never be told that a Douglas quitted his

There were no naval uniforms in Queen Anne's navy, but bargemen were often dressed in matching clothes, allegedly at the expense of the captain or admiral. These men are wearing red-visored caps, high-collared white shirts, short red trousers, grey leggings and black shoes.

post without orders,' and resolutely remained on board to burn with his ship.[1] On the other hand, the Dutch made good use of disaffected English pilots, and Pepys remarks in his diary that many Englishman were heard on board the Dutch ships speaking to one another in English and crying out that, 'We did heretofore fight for tickets; now we fight for dollars!'[2] On the 14th the Dutch succeeded in capturing the *Royal Charles* even though the British attempted to set fire to her twice, but both times the flames were quenched by the Dutch. The Dutch reached the sea with the loss of only two ships that had run aground and were set ablaze by the Dutch themselves, as well as eight fireships spent in the action.

In defence of the English, it should be said that a treaty between England and Holland was being negotiated at the time. The English assumed that the Dutch had no intention of sending their fleet to sea that year, but had not demanded a formal cessation of hostilities during the negotiations. With the connivance of France, the devious Dutch had arranged for the Queen mother to send a letter to King Charles II persuading him that if he sent his fleet to sea it might alarm the Dutch and French at a time when their thoughts were turned toward peace. Thus the English fleet remained moored in the river. Captain George Berkley expressed the English view of events when he remarked, many years later: 'The advantage to them was nothing, the expense considerable, and the infamy eternal.'[3] The Dutch view was understandably more

AN ADMIRAL'S BARGE c1710

Each bargeman is unique, and nuances in facial expression add charm and individuality. At least one rower appears to be distracted. The blades of the oars are decorated with stylised dolphins on both forward and rear surfaces.

sanguine. The successful raid on the Medway was considered a stupendous naval achievement, and the captured *Royal Charles* was brought to Hellevoetsluis, near Amsterdam, as a spoil of war. When the ship was broken up six years later, the splendid Royal Arms from the stern and the flag that once flew from the jackstaff were removed and stored in an Admiralty warehouse. One hundred years later they were displayed over the entrance to the model room in the Naval armoury.[4] Eventually they made their way to Amsterdam and the Rijksmuseum, where they were installed in 1885 and where they remain proudly displayed to this day.

The role of Admiralty ship models in the Dutch raid on the Medway emerged in the days following the raid, when the official inquiry began in an attempt to find a scapegoat for the disaster. Pepys records in his diary that on 13 June, with the Dutch raid still in full swing, blame was already being apportioned to, among others, the Office of the Ordnance for not providing Chatham and Upnor Castle with enough powder, the Navy board for not mooring the ships higher up river, and, of course, the Papists. Meanwhile, on that day, Pepys himself was busy dispatching his father and wife to the country with £1,300 in gold in their bags for safe keeping while he himself wore a girdle fitted with £300 of gold to carry awkwardly on his person. He also hastily made out a new will. But by then the worst was over, and the Dutch were reported to be sailing down river.

In the days that followed, Peter Pett (shipwright, model builder, and Commissioner of the Yard at Chatham) became the scapegoat for the whole affair. In the diary entry for 15 June, Pepys places blame on a want of boats 'that hath undone us'. Perhaps the boats could have been used to pull the ships up river, or to bring men to defend the ships, or to ferry troops to the fortifications. However, Pepys reports 'that they were employed by the men of the yard to carry away their goods; and I hear that Commissioner Pett will be found the first man that began to remove; he is much spoken against …'. Two days later, Commissioner Pett was imprisoned in the Tower of London. On the 19th, Pett was called into the Committee of the Council and charged with not carrying the great ships to safety, but instead using the boats to save his personal property, 'to which he answered very sillily', said Pepys. The silly answer is revealed several lines later. 'He said he used never a boat till they were all gone but one; and that was to carry away things of great value, and these were his models of ships; which, when the Council, some of them, had said they wished that the Dutch had had them instead of the King's ships, he answered, he did believe the Dutch would have made more advantage of the models than of the ships, and that the King had had greater loss thereby; this they all laughed at.'[5]

This episode speaks volumes about Pett's character, revealing him to be unsophisticated and hopelessly naïve. But he must have earnestly believed that the models were more valuable than the ships themselves, as no one imprisoned in the Tower, confronting hostile inquisitors, and facing an uncertain fate, would deliberately expose themselves to ridicule. This was undoubtedly the most highly prized collection of ship models ever assembled. Pett remained in the Tower for several months more, but as peace with the Dutch was finally achieved and interest in the Medway disaster faded, he was eventually released. This interesting footnote to the Second Anglo-Dutch War provides a unique insight into the enormous importance assigned to a collection of Admiralty Board ship models by at least one high-ranking seventeenth-century Navy official. Pett's culpability in the Medway disaster, however misplaced it may be, has been immortalised in a poem by Andrew Marvell that reads in part:

> *Pett, the sea-architect, in making ships,*
> *Was the first cause of all these naval slips,*
> *Had he not built, none of these faults had been;*
> *If no creation, there had been no sin:*
> *But his great crime, one boat away he sent,*
> *That lost our fleet, and did our flight prevent.*[6]

Graceful lines and colourful surfaces enhance the appeal of this twelve-oared, single-banked Georgian barge. The light construction can be judged by the scantling of the frames. The inboard decoration consists of fifteen gilt-framed painted panels along each bulwark positioned between the thwarts. The long, slender shape of this barge hints at swiftness when under way.

CHAPTER 23

A Georgian admiral's barge c1720

Acquisition

REMARKABLY, NO VESSEL WAS too modest to escape the attention of the Navy Board modellers. Barges such as the one represented by this model must have been very common in the early eighteenth century, and one wonders if there was any motivation other than to create a lovely object that led to the building of this little model of a twelve-oared, single-banked, ship's barge. We purchased this model at auction at Sotheby's in 1994.

Provenance

This lovely little model appeared at a Sotheby's auction in London on 11 May 1994. The identity of the vendor, where it came from and through whose hands it passed remain a mystery.

Description

CONDITION

This model was in excellent condition, though lacking a rudder, sternboard and the aft ends of the gunwales. Six segments of the gangboard were also missing. These parts were replaced by Philip Wride, who also carved the dolphin cradles upon which the model now rests. The wonderful polychrome decoration is original. When the model was purchased, there was no case. Nonetheless, it must have been protected in a storage box of some kind, or else it could not have survived relatively intact for so many years.

CONSTRUCTION

Scale: 1/24 Hull length: 19in

This is a contemporary Admiralty Board model of a twelve-oared barge built in the time of King George. It is an example of the type of barge carried on board a ship for use by a flag officer. The open-frame construction is of boxwood, but has been completely over-painted in a taupe colour. The model is quite colourful. The inside bulwarks are red and are decorated with a series of twelve gilt-edged rectangular panels positioned between the thwarts, each painted with leafy foliage in shades of grey and white. An additional three panels line the sides of the sternsheets, and a similar decorated lunette panel adorns the backboard. Interestingly, there are additional small panels lining the bulwarks in the coxswain's dickey, but these are not decorated and are simply painted red. The gunwales are also decorated outboard with sinuous leafy decoration in grey and white in a gilt-edged panel running fore and aft the length of the barge.

The barge is carvel-built and has twelve thwarts for single-banked oars. The centre thwart is wider than the others and has a hole bored to accommodate a mast. Lockers faced with

The first thwart abuts the breast hook, which has a hole bored to support a jackstaff. A lifting ring is fitted to the inside of the stempost.

rectangular moulded panels are fitted below the sternsheet benches with round openings at their forward ends. There are seats to both port and starboard for the coxswain. The keel is rabbeted to receive planking and a keelson has been fitted. The stem and sternpost are joined to the keel with plain scarphs. The sternpost is raked and backed by a false-post and a deadwood knee. The stempost terminates in a shaped bracket and is faced by a metal strip decorated with beading that reaches to just below the second strake. Fifteen frames are fitted, and the aftmost frame forms the fashion piece and is raked and fastened to the sternpost. Strips of footwaling are fitted along the floor and extend the length of the barge. Two planks are fitted below the gunwales, secured by brass pins. The gunwales are fitted with twelve rowlocks of standard type. The rudder is hung on the sternpost by two gudgeons and pintles. As usual, the upper pintle faces downward and is attached to the rudder while the lower, longer one, faces upward and is attached to the sternpost.

Historical Perspective

THE ELUSIVE CRAFTSMEN WHO CONVERTED WARSHIPS INTO WORKS OF ART

The existence of an Admiralty Board model of such a humble craft as this raises questions about the pervasiveness of this unique model-making practice. It is remarkable that despite the obvious time, effort and expense involved in making Admiralty Board models and the evidence that production continued for over 150 years, there is almost nothing known of the men engaged in this enterprise.

We are reminded that in the seventeenth and eighteenth centuries, craftsmen took pride in beautifying even the humblest object. In the shipbuilding industry, artistry in building models as well as ships must have been assumed, and no special notice was taken when a fine result was achieved. Nonetheless, based on observations made over the past thirty years, we have arrived at some conclusions about the men who made these models. In the seventeenth century it appears that model building was part of the job description of a master shipwright. In his autobiography, Phineas Pett[1] proudly describes how he made models for almost all of the ships he proposed to build, often before they were commissioned. In 1596 he made a small model for William Cecil, Lord Burghley, who was the Lord Treasurer. In 1599 he made another small model 'very exquisitely set out and rigged'. In 1607 he made a model for the Lord High Admiral 'most fairly garnished with carving and painting', which was subsequently shown to a delighted King Charles I. This model was the basis for the *Prince Royal* of 1610. In 1634 Pett showed another of his models to the King. This was the design for a revolutionary 100-gun ship that was to become the *Sovereign of the Seas,* which, when launched in 1637, had the distinction of being the largest and most costly ship of any built up to that time. The model was evidently kept at the home of Phineas Pett's son, Peter, who was charged with building the ship, and it was seen there by Peter Munday, who described it as 'of admirable workemanshipp, curiouslye painted and guilte, with azur and gold, soe contrived that every timber in her might be seene and left open and unplanked for that purpose, very neate and delightsome'.[2] This description confirms that as early as 1634, the Admiralty Board model style of open-frame construction was already in use.

The first official mention of modelling as part of the shipbuilding enterprise is in the form of an official order from the Navy Office dated 2 December 1645:

> It is ordered that the Comm^{rs} of the Navy doe conferre wth the Master Shipwright and his assistant appointed to buyld the Three Ffrigatts and doe cause them to frame a platt or moddle of them severally and present the same to the Lords and others of the Committee of the Admiralty wth all convenient speed. –Giles Green.[3]

The shipwright in question was Peter Pett, master shipwright of Woolwich and Deptford yards, and the three ships were the *Assurance, Nonsuch* and *Adventure*. The next official reference we know is dated 12 April 1649:

> The council of State by their order of the 29 March last ordering that five ships be built … But before the said builders proceed in building, this Cimmittee desire you to order the builders to present models of the frigotts they severally undertake, according to the direction aforesaid … Your very loving friends, H. Vane and Valentine Wanton.[4]

The cubby for the coxswain is visible, and interestingly the decorative panels along the bulwarks are unpainted in this compartment. A series of holes bored into the gunwales provided support for a removable awning frame.

These orders leave open the possibility that the term 'model' refers to a two-dimensional plan, or drawing, but given that more than 100 seventeenth-century models have survived, and only a handful of drawings, it is likely that regardless of whether drawings were submitted, the Navy Board committees were also presented with three-dimensional models. Confirmation of this practice comes from a note by the Frenchman Tourville, who wrote in 1686, 'They never build a ship in England without first making a model so as to find out faults and correct them easily.'[5]

Samuel Pepys provides more confirmation that in the seventeenth century, master shipwrights were also model makers. Pepys records that Christopher Pett gave a model to Mr Coventry,[6] and that Anthony Deane gave a model to Pepys, in addition to making and bequeathing a model of the *Royal James* to Christ Hospital Mathematical School (see Chapter 1). Pepys, in his *Diary*, also relates that during the Medway disaster of 1667, Peter Pett carried his own collection of models to safety while leaving the ships themselves to the Dutch, as related in Chapter 22. Sutherland, a shipwright best known for writing two books on shipbuilding, was also a model maker. That model making enjoyed a certain status may be inferred from the observation that Queen Anne's husband, Prince George, made ship models and had a special workshop created for this purpose,[7] and that even Peter the Great of Russia is said to have tried his hand at model building during his visit to English and Dutch shipyards.[8]

In the seventeenth century, it was not uncommon for some builders to sign and date the wooden supporters for their models. The Earl of Pembroke has a 3rd rate that bears the initials 'JS' and the year '92' carved on the dolphin cradles. The initials most likely refer to John Shish, master shipwright, and one member of a family of shipwrights, and the model bears a striking resemblance to the *Hampton Court*, a ship built by Shish in 1678. It has been speculated that the date on the cradle refers to the date of completion of the model, not the ship,[9] and it cannot be claimed with certainty that Shish built the model as he may have only built the ship it represents. The model of the *Boyne*, a 3rd rate of 1692, has the name of the ship's builder carved prominently on the decoration at the break of the poop deck, 'YE BOYNE Bt BY MR HARDING DEP …' It is satisfying that the *Boyne* was indeed built by Fisher Harding at Deptford. But again it is not certain that Harding built the model.

We think that a strong case can be made that Fisher Harding was a model maker because there are two other models that have his initials carved into them. One is a model at the NMM of a 3rd rate, c1698, with the initials 'FH' and the date '1698' carved into the dolphin supports. The model, as is often the case, does not match any known ship of the same date. Most interesting of all is the model of the *St Albans* of 1687, at Trinity House, London. The *St Albans* was built by John Shish at Deptford, but the model has the arms of Fisher Harding at the break of the quarterdeck. Harding was working at the Deptford dockyard at the time,[10] and one suspects that he may well have made the model and signed it. Of interest, this model and the model of the *Boyne*, which also has Harding's initials carved in the decoration, are both constructed with an unusual open framing style where every other floor and futtock is omitted. This style of model building appears to characterise Harding's work. It is notable that the crowning achievement of Harding's career was his magnificent 1st rate, the *Royal Sovereign* of 1701. A model of this ship survives in the Royal Naval Museum in St Petersburg, Russia, and, interestingly, it is framed in the unusual open framing style of the *St Albans* and the *Boyne*, which we suspect is a hallmark of models made by Harding.

There are a number of other models dating from the end of the seventeenth and beginning of the eighteenth century that have the builder's initials carved into the cradles or the model itself. Two examples are in the collection of the Pitt Rivers Museum, Oxford. The model of the *Lizard*, a 6th rate of 1697, has the initials 'RS' along with the date '97' carved into the dolphin cradles. The master shipwright Robert Shortis built the *Lizard* at Sheerness dockyard and also presumably made the model. Another model in the Pitt Rivers Museum, of a galley frigate, also has initials and a date carved in the cradles. In this example, the initials are 'IE' or possibly 'JE' and the date is '1702'. The model is very much like the *Charles Galley*, but that ship was rebuilt in 1693 and no evidence has come to light suggesting another rebuild around 1702, nor is there a shipwright at the time with matching initials. A similar mystery surrounds a model of a 4th rate at the National Maritime Museum, Greenwich, with the initials 'IL' or 'JL' and the date '1701' painted in a cartouche on the stern. The initials could fit either of two master shipwrights, Mr John Lock of Plymouth or Joseph Lawrence of Woolwich, but there is again no match for a ship of the date and size of the model. The Pitt Rivers Museum also has a model of a 3rd rate of 1706 that was originally given to the Ashmolean Museum, Oxford, by Dr George Clarke, who was as a Lord of the Admiralty in 1710–14. The Book of Benefactors records that the model was made by William Lee Esquire; presumably this was the same William Lee who was master shipwright, and subsequently surveyor, at Woolwich dockyard.

A new page on the identity of model shipwrights was written in February 1992, when Angus White of Horsham, England, discovered a note inside the model of the *Leopard*, a 50-gun ship of 1790 belonging to him. The note reads: 'This moddle was made by Geº Stockwell at Sheernefs in the year of our Lord 1787 in the 56 year of his aige.'[11] A model of the *Bristol*, a 3rd rate of 1775, was sold at Sotheby's in 1991, and Angus, struck by the resemblance of this model to his own of the *Leopard*, persuaded the buyer to look inside. Astoundingly, a note was found in this model that reads: 'This model was made May the 7 1774 By Geº Stockwell Shipwright at Sheerness Yard.'[12] Angus White has uncovered additional details concerning Stockwell's shipbuilding career, including the fact that he began his apprenticeship at Sheerness at age fourteen, was entered as a shipwright at Sheerness age twenty-one, and by the time he died, aged sixty-five, he had served forty-three years and eight months in the service.[13] We have concrete evidence, attested in his own hand, that George Stockwell, shipwright, built Admiralty Board ship models, but the record is mute on whether he had a hand in building the ships. Possibly he was employed strictly as a model maker. The models he made were all fully planked, a style that had supplanted the open-frame mode of construction of the seventeenth and early eighteenth centuries, and therefore could easily conceal a note. How many other models by Stockwell are still unidentified in museum collections? There are at least half a dozen candidates in museums in the US and the UK, and it is almost certain that more notes would come to light if they were all inspected.

By the second half of the eighteenth century, there were enough models distributed in homes and enough interest in them to support a secondary commercial activity devoted to the care as well as construction of model ships. This is attested to by an advertisement in the form of a trade card belonging to Allen Hunt. The first example of this card to come to light is one in our collection that was attached to the mahogany baseboard of the *Lion* model. It was photographed by the Science Museum in London in the 1930s, and this photograph has been widely reproduced, and an enlarged poster-size version is on display at the NMM.[14] The text of the card reads:

> Models of Ships, Cutters & Boats Built by Draft & Scale in the most accurate manner, by ALLn HUNT, Ship Modeller, No.7, King's Row, Horsleydown, Southwark, Models Kept & Cleaned

Alongside the text is an engraved image of the portly craftsman himself, along with a selection of models that clearly date to the second half of the eighteenth century. It is interesting to note that this trade card went through at least two printings. The example cited above is in brown ink, but several years ago we obtained a second one from a rare book dealer in London, and this one is in black ink. Allen Hunt was born around 1742 and could not have made the *Lion* model. It is likely that he cleaned it and may have replaced the original supports with the cradles it now rests upon. This suggestion is based upon the existence at the NMM of an

A GEORGIAN ADMIRAL'S BARGE c1720

identical set of cradles also attached to a mahogany baseboard, which supports a 1719 version of the *Royal* William.[15] The style and workmanship on this model is quite different from the *Lion* and could not be by the same builder, yet the cradles and baseboards clearly are. Such baseboards are not appropriate for the date of either model, and we believe that they may represent the handiwork of Mr Hunt.

The lapstrake construction of this carvel-built barge is evident from the two planks fitted below the gunwales. The stempost is faced with a metal strip decorated with beading, and this is continued along the upper surface of the shaped stempost termination.

This is a remarkably complete model presenting a miniature vignette of activity aboard a Royal Navy barge. The masts, spars, and reefed sails included here are rarely seen on barge models.

CHAPTER 24

An admiral's barge c1775

Acquisition

WE ACQUIRED THIS MODEL at an auction of the marine collection of the Museum of the Citadelle of Vauban on Belle Île, France, in July 2010.

Provenance

This model was collected by André and Anna Larquetoux and displayed at their private museum in the Citadelle of Vauban until 2010. Its earlier history is sadly unknown to us.

Description

CONDITION

The model is complete and original and includes ten crew members, a captain and a coxswain, along with ten original sweeps and two masts bearing yards and reefed sails. The barge crew have hair on their heads and shirts and trousers made of fabric that is very fragile and shows signs of deterioration. Similarly, there are finely quilted seat cushions in the sternsheets that are also very fragile. The barge, sweeps, and figures retain delicately painted decoration that is in a wonderful state of preservation.

CONSTRUCTION

Scale: 1/24 Length: 17½in Height: 10in

This is a contemporary Admiralty Board model of a carvel-built, ten-oared, single-banked admiral's barge complete with captain, helmsman and oarsmen. Nine of the oarsmen are at their seats, with articulated arms that can grasp the sweeps. The tenth oarsman stands at the bow with a boat hook in hand as if prepared to grapple a mooring rope. All the sailors are dressed identically with sewn fabric shirts, trousers and broad belts, as well as neck scarves. They all have realistic shoulder-length hair and wear carved wooden helmets painted with the GR cipher of King George III. The captain and helmsman are static fully carved wooden figures. The captain sits in the sternsheet with his right hand tucked into his shirt and his legs crossed. The helmsman steers the boat from a standing position with the tiller held behind his back. The captain and helmsman both wear blue jackets, ruffled shirts, breeches, stockings and gold-buckled shoes. The captain's jacket has gold buttons and lacing as well as gold-decorated cuffs and edging on his tricorn hat. The helmsman wears his hair in a pigtail or queue.

The external hull is painted white, with an orange sheer strake and a decorated gunwale. The fore edge of the stempost is

One of the ten oarsman stands in readiness at the bow as if preparing to secure a mooring line. Note the striped ticken trousers worn by all the barge crew.

HISTORIC SHIP MODELS

Although commonly referred to as an admiral's barge, the ship's captain was usually the highest-ranking officer on board. A captain wearing the dress uniform for 1774 is shown seated in the sternsheets of this barge. Note the quilted seating cushions.

protected by a metal strip extending to the heel of the stem that is faced with beading above the waterline. The painted gunwale decoration includes dolphins, fish and waterfowl along with the more typical intertwining acanthus leaves. There are finely painted decorative panels inside the bulwarks, some of which depict masks and military flags against backgrounds of floral decoration. The sternsheet benches are topped by quilted cushions fashioned of gold-coloured fabric. The ten sweeps are painted green with dolphins painted on the aft surface of the blades. There are two erected masts, one amidships and the other at the bow, both fitted with yards and reefed sails. A gangboard runs between the mast-bearing thwarts.

Exhibitions

Belle Île, France, Museum of the Citadelle of Vauban, prior to 2010.

Historical Perspective

UNIFORMS INFORM THE COLLECTOR

Remarkably, sailors did not have regulation uniforms until 1857. Nonetheless, the crews of Royal Navy ships in the eighteenth and nineteenth centuries generally wore similar clothes. For example, sailors throughout this period typically wore striped ticken trousers, but depictions of sailors' clothes of the late eighteenth century are quite rare. Interestingly, accurately made miniature striped ticken trousers are worn by the sailors on this model.[1] The caps worn by barge crews were not regulation, but were usually furnished by the commanding officers and were often personalised. The maker of this barge has kept the clothing generic, and rather than display the name of a specific ship or commander, the caps bear the cipher of the reigning sovereign, King George III. The dress of the commanding officer and the helmsman are more informative. Uniform regulations for Royal Navy officers were first issued in 1748, and uniforms were very specific. The coat worn by the officer in the sternsheets conforms exactly in terms of cut, collar, facings and cuffs, to the captain's dress uniform for 1774, so that the barge can date from no earlier. Interestingly, he wears blue breeches, which were in style from 1748 until 1767, but were definitely out of fashion after 1767, when the Admiralty stipulated white breeches were to be worn. It is therefore unlikely that the barge dates from much later than 1774, and even then our captain would appear to be a bit old-fashioned. The rather prim helmsman is a warrant officer, probably a master, as they typically wore simple blue coats.

The sailors' clothing and hair add an air of realism to the barge crew. Painted dolphins decorate the oar blades.

AN ADMIRAL'S BARGE c1775

Above: The helmsman's plain blue coat marks him as a warrant officer, most likely the ship's master. He wears his hair in a characteristic pigtail or queue. The GR cipher of King George III is displayed on the barge caps worn by all the oarsmen.

Right: The seating arrangement of this single-banked barge is very evident in this view. Decorative panels line the bulwarks inboard.

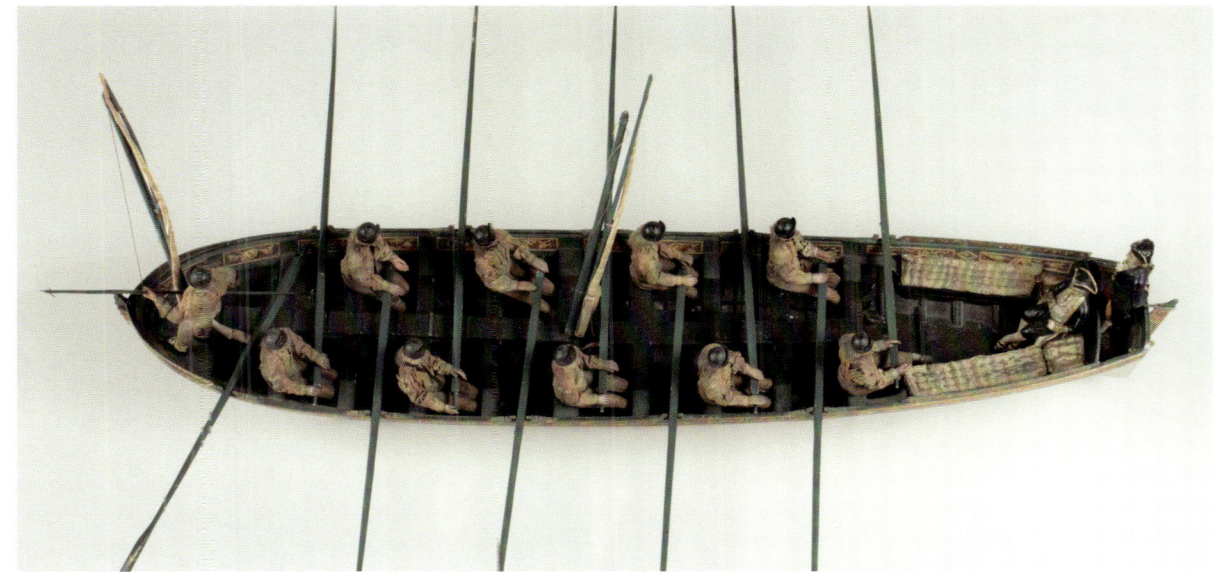

There are fifty-seven hand-carved and painted figures occupying and manning the barge, with not an inch of wasted space.

CHAPTER 25

A troop transport c1810

~ Acquisition ~

WE PURCHASED THIS MODEL from the descendants of Clarkson Collins Junior in 1992 while it was on view at the Mystic Seaport Museum. Years later we reluctantly traded it away in order to acquire a model of an admiral's barge from the time of Queen Anne (see Chapter 20). Happily, we were able to reacquire the model in 2009.

~ Provenance ~

This model was part of the Clarkson Collins Junior collection and was loaned to the Boston Museum of Fine Arts, where it was on view in the ship model room from 26 February 1929 until 18 March 1931. It was subsequently placed on long-term loan to the Mystic Seaport Museum until 1992, when we purchased it from Mr Collins' descendants.

~ Description ~

CONDITION

The model is in original condition, equipped with sixteen Navy oarsmen and one naval officer, as well as a complement of thirty-eight Army troops, one sergeant and two musicians. All the figures, fittings and the transport itself are original and complete, including their polychrome decoration. The brass support stanchions are contemporary to the model; only the walnut-veneered plinth is modern.

CONSTRUCTION

Scale: 1/24

This is a model of a double-banked, sixteen-oared troop transport of the Napoleonic War period. The model includes forty-one soldiers belonging to a centre company of a battalion from the 3rd Foot Guards Regiment. The miniature sailors and soldiers are carved out of wood, but their hands are made of a moulded, hardened paste. Each figure is secured to his assigned place by a wooden peg fitted to a hole drilled in the bench seat. The oarsmen are wearing blue waistcoats, white trousers, and black handkerchiefs and hats. A naval officer mans the helm in the sternsheets, seated alongside the sergeant who has his drawn sabre held upright in his right hand. Two uniformed foot soldiers flank the officers to port and starboard, and three sit facing them. Fourteen more troops are seated facing each other along fore and aft benches between the stern and mast step. The soldiers sit shoulder to shoulder with bayoneted muskets clasped before them. Thirteen additional troops

All the uniforms are accurately detailed, which allows them to be identified and dated, placing the origin of the model within the narrow range of 1806–12.

The Army troops wear redcoats, while the oarsmen and helmsman sport the bluecoats of the Royal Navy. The Army troops all carry knapsacks and canteens.

sit facing each other on the benches between the mast step and the foremost thwart in the bow. Four more 'redcoats' are seated precariously on the bulwarks at the bow. A fifer and drummer are included playing their instruments – one can easily imagine the strains of a marshal tune – and are seated on the port fore-to-aft bench just before the mast step.

A reinforced thwart drilled to accept a mast lies amidships, secured by a pair of hanging knees bolted to the gunwales. The bow thwart is pierced to accept a small mast and is also secured to the gunwales by small hanging knees. With the exception of the thwarts amidships and at the bow that extend side-to-side to the gunwales, the remainder of the oarsmen's benches are shortened to allow space for the troops to sit centrally. The broad-beam, flat-bottom hull is of lapstrake construction with each strake fixed to the frames by brass pins. The sheer strake is painted in a cream colour, and the top strake is black.

Exhibitions

Boston, Museum of Fine Arts, February 1929–18 March 1931.
Mystic, Connecticut, Mystic Seaport Museum, 1942–92.

Historical Perspective

REDCOATS AT SEA

We know of two nearly identical models of troop transports from the period of the Napoleonic Wars, or more appropriately, the Wars of the French Empire, 1803–15. One is in the Thomson Collection at the Art Gallery of Ontario, and the other is this example. The boats themselves are identical; both have equal numbers of sailors and troops including a piper and drummer. The present example is slightly older, and the detailed sculpting and painting of the soldiers allows them to be identified fairly accurately.[1] The troops are Army infantry foot soldiers. They are wearing 'stovepipe' shako hats that taper slightly toward the top, a style authorised in October 1806. The drummer and fifer are wearing blue jackets with red facings, which is just the opposite of the troops, who are wearing the eponymous red jackets with blue facings. This was standard practice in the Army until 1812, when drummers and fifers reverted back to wearing the same regimental colours as everyone else. The model must therefore date between 1806 and 1812.

The soldiers are decorated with sufficient detail to also allow identification of their unit. The dark blue colour of the facings (lapels and cuffs) worn by the soldiers identifies them as members of one of only three regiments of Royal or Foot Guards. These were

The fifer and drummer are depicted in the midst of playing their instruments and it requires little imagination to bring the whole scene to life with image and sound.

A TROOP TRANSPORT c1810

The lapstrake hull is finely made and represents the culmination of a long evolution in the design of troop transports for the Royal Navy.

For some reason, large numbers of troop transport models were made. This is a second earlier example in our collection dating from the period of the Seven Years' War. It originally held a troop of grenadiers. Unfortunately, only a few survive, along with several oarsmen. This mid-eighteenth century model has a square stern.

among the most famous of all the Army regiments, each with a long and distinguished history. The number 3 painted on the troops' knapsacks indicates that they belong to the 3rd Regiment of Foot Guards – an elite unit that was raised by Archibald 1st Marquess of Argyll in 1642 and continued to serve with distinction, particularly during the Napoleonic Wars. The fact that the woollen plumes or tufts worn at the front centre of the troops' stovepipe shakos are coloured white over red identifies them as belonging to a 'centre' infantry company. Therefore, the soldiers being transported belong to a centre company in a battalion from the 3rd Foot Guards Regiment.

What can we tell of the Navy officers in the sternsheets?[1] The helmsman wears a black, unadorned, two-cornered cocked hat positioned fore and aft. This bicorn style replaced the earlier tricorn hat around 1795, when uniform regulations were issued, and most officers wore them fore and aft, while admirals wore them athwartships. The uniform regulations also removed the white lapels of officers above the rank of lieutenant, and our helmsman's coat is dark blue with a high collar and blue lapels, as appropriate for a date after 1795. The hat and coat therefore tell us he dates from after 1795. The absence of epaulettes or lace at his cuffs tells us his rank. Admirals, captains and commanders all wore epaulettes of varying number, but lieutenants wore no epaulettes at all. Beginning in 1812, however, white lapels were reintroduced for all officers and for the first time lieutenants received a single epaulette on the right shoulder. Our helmsman is therefore a lieutenant dating from after 1795 but no later than 1812. Beginning in 1805, officers in the Royal Navy carried a regulation pattern sword in place of the cutlass of earlier times. The regulation sword had a gilt stirrup guard, a lion's head pommel, ivory grip and a long straight blade carried in a black leather scabbard, exactly like the sword carried at the lieutenant's waist. The hat, lace and cuffs, sword, blue coat with blue lapels and no epaulettes identify the helmsman as a Royal Navy lieutenant in 1805–12. These dates conform very nicely with those based on the Army uniforms.

SCOTS GUARDS

The Scots Guards, to whom our model troops belong, played a prominent and illustrious role in the Napoleonic Wars. Battalions of the 3rd Foot Guards contributed to the victories at Copenhagen in 1807, and played a decisive role in many of the battles of the Peninsular War in Spain and Portugal. They also distinguished themselves at the Battle of Waterloo on 18 June 1815, where Napoleon was finally and decisively defeated.

All of the landing craft models feature lapstrake hulls made of boxwood. These light-weight strong boats were well designed for use along the western Atlantic coast.

CHAPTER 26

A ship's boat c1750

Acquisition

THERE IS BEAUTY IN the lovely curves and airy lightness of this humble boat. It is nonetheless a wonder that the Navy Board modellers should have bothered to create this little model. The open-frame construction lends a deceptive delicacy to what must have been a very durable work boat. We purchased this model at Sotheby's, London, in 1987.

Provenance

This fine dockyard model of a relatively humble longboat appeared for sale at a Sotheby's auction.

Description

CONDITION

When we acquired this model, it was in original condition, but unrigged. However, it had fittings for rigging, including ivory deadeyes and pulley sheaves, and we decided to add the mast, bowsprit and rigging to present a more complete portrayal of this little craft. The rigging did not involve any alteration to the original model and is reversible.

CONSTRUCTION

Scale: 1/48 Hull length: 9½in

Left: This little boxwood model represents a ten-oared, single-banked longboat. Warships of all rates carried similar boats on board, and the basic design changed little throughout most of the eighteenth century. This example is shown with a mast stepped and a gaff rig.

This is a delicate open-frame model of a ten-oared, single-banked, 36ft longboat or ship's launch. The hull is carvel-built, with twenty-four frames, the fifth through twentieth of which are shown with lower futtocks that overlap the floors and are held together by fine brass pins. There is a windlass amidships equipped with windlass bars, and the overlying thwart is removable to allow the windlass to be worked. There are ten thwarts let into notches in the rising, each with a finely incised decorative edging. The first and fourth are braced by knees to support the jib-boom and mast respectively, and the fifth is removable to accommodate the windlass bars. A metal mast clamp is fitted to the mast thwart, as well as belaying pins either side of the mast. The thole pins, deadeyes and pulley sheaves are all made of ivory. A keelson and four bottom boards are fitted from the third frame to the twentieth, neatly pinned to the frames with brass pins. The pins were presumably driven through pre-drilled holes, since none of the delicate frames or planks are split, and were then filed off flush inboard and out. The sternsheet footwaling is also fitted with the planks represented by two shaped panels. The sheer strake is painted red and bordered by a shaped moulding, and one plank is displayed below the gunwale, fastened to each frame by a double row of brass pins. The stem and sternpost are rabbeted to receive planking, and the stempost is backed by a stemson and deadwood. The stem and sternpost are fastened to the keel with scarph joints, and the garboard strake is neatly pinned in place by two pins driven into each frame. The transverse sternsheet bench is fitted as a locker with a hinged seat secured with finely made miniature butterfly hinges. On the port side of the stem there is a collar to take the jib-boom, and sheaves fashioned of ivory are fitted at the bow to the starboard side of the stem and to the stern. These sheaves, together with the windlass, were used to haul the anchor buoy rope, one of the primary tasks of the launch. The model is shown with the single-mast gaff rig common on launches of the eighteenth century.

At the NMM there are two models of ship launches of the mid-eighteenth century, one rigged and one not, and both are very similar to this one. One difference, however, is that both of the

The sternsheet bench seat is hinged and opens to provide access to a locker. The sternsheet footwaling is flat and at a higher level than the footwaling below the thwarts, which is pinned directly to the frames and follows their curves.

NMM examples are double-banked, while ours is single-banked. Perhaps these little models were made to demonstrate small differences like this.

Historical Perspective

A FEAT OF NAVIGATION SUCH AS THE WORLD HAD NEVER SEEN

Lieutenant William Bligh is known to modern readers as the captain of the *Bounty* and instigator of the most famous mutiny in naval history. However, to his contemporaries, he was more famous for his feat of navigation in piloting an overloaded ship's boat nearly 4,000 miles across the open ocean, safely conveying all eighteen castaways on board to safety. This achievement was hailed as the greatest boat journey ever undertaken, and the little vessel captained by Bligh was very much like the one represented by our little model, only smaller.

The boat journey began on 29 April 1789, when Fletcher Christian forced a bound Captain Bligh over the side of the *Bounty* and into the ship's launch, a boat already so heavily laden with provisions, as well as eighteen men, that she rode with her topsides only 7in above the calm waters. So overburdened was she, that additional loyal officers and men chose to remain with the mutineers rather than risk near certain death by joining the boat's company. The six-oared launch was only 23ft in length and was stored with provisions that would normally support nineteen men for only five days.

Before setting off across the open ocean, Bligh landed on the island of Tofoa to secure more food and water for the journey. The natives attacked them as they were departing, and John Norton, quartermaster, was killed. Forty-eight days and 3,618 miles later, Bligh landed the little boat on a beach in Timor after a passage punctuated by Pacific gales, drenching rains, numbing cold, thirst, starvation, exposure and exhaustion, but with no additional loss of life. This was a stupendous feat of navigation, accomplished solely with the aid of a compass, a quadrant and miscellaneous navigation tables. The launch was rigged with two sails and provided with oars, and in all but the calmest seas, most of the men were kept busy bailing. Bligh ruled with supreme authority. He established the rules and routines the men lived by, measured out the rations of food and drink, divided the men into watches, took soundings and sightings, maintained a ship's log and generally succeeded better as captain of the launch than he did as captain of a King's ship. Not surprisingly, when Bligh came to write his memoir of the eventful voyage of the *Bounty*, entitled: *Narrative of the Mutiny on Board His Majesty's Ship Bounty; and the Subsequent Voyage of Part of the Crew, in the Ship's Boat, From Tofoa, one of the Friendly Islands, to Timor, a Dutch Settlement in the East Indies*, he described the mutiny briefly, and devoted most of the text to a detailed narrative of his exploits on board the launch.[1]

The first and fourth thwarts are braced by knees to help support the stresses associated with the jib-boom and mast respectively. A metal stanchion secured to the first thwart supports the aft end of the jib-boom, and a metal collar secured to the stempost provides additional support.

A SHIP'S BOAT c1750

Eventually, ten of the mutineers were captured on Tahiti and brought back to England for trial. Three were sentenced to death and were hung upon the yardarms of the *Brunswick*, an execution witnessed by the crews of every ship in Portsmouth harbour. Bligh was later made Governor of New South Wales, but in a telling replication of previous events, the colonists rebelled under his rule, the British troops mutinied and Bligh was imprisoned for two years before being sent back to England. Nonetheless, Bligh eventually attained the rank of vice admiral and died in London in 1817 aged sixty-four. There were many mysterious turns of fate in the eventful life of William Bligh and much that remains contradictory and ambiguous to this day. But one can safely conclude that Bligh and his crewmates owed their survival not only to his navigational skills, as is widely acknowledged, but also to the sailing qualities of the sturdy little ship's boat that served them so well.

References

Bligh, William and Christian, Edward, *The Bounty Mutiny* (New York: Penguin Books, 2001).

Alexander, Caroline, *The Bounty* (New York: Penguin Books, 2003).

This ship's launch was a utilitarian vessel, often used to help haul the anchor buoys. It was equipped with a windlass amidships with a removable overlying thwart. A metal mast clamp is fitted to the mast thwart with belaying pins either side of the mast. The thole pins, deadeyes and pulley sheaves are made of ivory or bone.

218

CHAPTER 27
A Dutch state yacht c1690

~ Acquisition ~

THIS MODEL FIRST CAME to our attention when it was sold by Sotheby's, Mak van Waay, Amsterdam, in October 1975. It was lot #281, correctly described as a period model of a seventeenth-century Dutch state yacht. Though we very much liked the model, we were medical students at the time and it was too expensive for us. Twenty-five years later, however, our circumstances had changed and, surprisingly, we had another chance.

The same model appeared as lot #62 in a sale at Christie's, Amsterdam, on 12 September 2000. This time around it was described as a nineteenth-century model and had a much lower estimate. The model had evidently come to the auction house without any information. Under the circumstances, Christie's took a conservative approach and missed the age of the model by one or two centuries, giving us an opportunity to get a bargain as long as no one else recognised the mistake. As it happened, no one else did.

~ Provenance ~

The model appeared in Amsterdam in 1975, but its prior history and original owner are unknown.

~ Description ~

CONDITION

The model has survived very nearly intact. The topsail and some rigging lines are old replacements, but the remainder of the sails and rigging appear to be original. There is evidence of old touch-up and infill on painted surfaces where the original pigment has flaked off. All the structural elements are original, including the rudder, leeboards, anchors, cannon and decoration. The cradles and plinth are modern.

CONSTRUCTION

Length overall 38.5in, Height 40in

This is a contemporary model of a Dutch state yacht dating from the end of the seventeenth century. The hull is built up plank on frame with wooden planks and strakes held in place by metal pins. The main wale is studded with round-headed fastenings. The wood used for the hull planking appears to be a species of pine. Leeboards are fitted and are constructed, hinged and rigged as on

The carving of a hunter dominates the taffrail, flanked by roundels depicting the Zeeland arms to port, and the red castle of Aardenburg to starboard. The transom is painted with the prospect of a Dutch town viewed from the sea, presumably the Zeeland town of Aardenburg.

Unlike British dockyard models, Dutch models are typically fully planked and rigged with sails, as in this example. The shallow draft and leeboards allow these vessels to negotiate the inland waterways of the Netherlands.

the full-size vessel. The model is fully rigged with a single mast and fitted with a suit of hand-stitched linen sails consisting of one square sail, a loose-footed gaff and two head sails, all equipped with luffs, leeches and grommets. A wortel is perched on the mast head and a flagstaff at the peak of the gaff. The anchor cables pass through hawse holes faced by decorative carvings in the form of grotesque masks. The cables then pass around the windlass and are led below a hatch cover to the cable hold below. A pawl that serves to check backward motion of the windlass is hinged to the aft side of the stem and passes beneath a pin rail to engage a toothed wheel let into the windlass barrel. Hatches on either side of the mast lead to the crew bunks and the caboose or galley. The yacht is armed with

The emblematic lion figurehead holds a book in his forepaws, with the red castle tower, a symbol of the town of Aardenburg, and a dolphin depicted in the trailboard.

eight 3pdr cannon. The octagonal glass-sided lantern lets light into the 'long-room' below, equipped with fore and aft wooden benches on either side. A stairway descends to the captain's cabin. At the stern two glazed doors lead into the grandest room on the vessel, the pavilion, which is furnished with a central octagonal pedestal table. Benches are incorporated into the quarter galleries. An iron tiller with a cranked up end passes below the pavilion deck and can be swept side to side by a helmsman standing on the main deck.

Exhibitions

Washington DC, National Gallery of Art, *Water, Wind, and Waves: Marine Paintings From the Dutch Golden Age*, 1 July–21 November 2018.

Washington DC, National Gallery of Art, December 2018–present.

Historical Perspective

THE HUNTER OF AARDENBURG

The decorations on this model are replete with symbolic significance that was likely very readable to a seventeenth-century Dutchman, but much more difficult to decipher in the twenty-first century. The figurehead depicts a gilded rampant lion holding a book. We have not been able to find this symbol associated with Dutch heraldry. Possibly the figurehead may be an emblem.[1] The use of personal emblems was widespread among the gentry in Europe in the seventeenth century, and emblem books were published to help individuals choose appropriate personal symbols. An emblem of a lion holding an open book facing the viewer appears as Emblem No. 67 in a book of emblems by Julius Zincgreff published in 1619.[2] The emblem answers the question of which virtues a governor or sovereign should have; he should have the strength of a lion tempered by prudence and wisdom. This emblematic lion may have a dual role and may also represent the golden lion of Nassau. William III of Orange-Nassau was stadtholder of Zeeland from 1672 to 1702, as well as serving as King of England, and the golden lion may be a reference to him.

The trailboard carving incorporates a red castle tower emblematic of the town of Aardenburg, a small coastal town in the province of Zeeland. Its medieval name was Rodenburgh (Red Castle), and it is the oldest town in Zeeland. Behind the tower is a dolphin. The pavilion is decorated outboard with friezes depicting lions, dancing cherubs and female symbols of bounty. There are spirited carvings of boys riding dolphins just below the pavilion lights. The stern decoration is dominated by a centrally placed carving of a hunter. He is dressed in a short-belted coat, trousers, boots and a wide cape, typical dress for a hunter in the baroque period, and he has what may be a ferret at his feet. Ferrets were common hunting companions at this time. His forearms are missing, but there appears to be a musket over his shoulder. A bag tucked in his belt may be to store his booty, suggesting he may be hunting fowl. Both quarterfigures (*hoekmannen* in Dutch), are dressed like the central figure and are meant to be hunters as well. All three figures are bearded. Flanking the central carving are the arms of Zeeland and Aardenburg. The Zeeland coat of arms are on the port side, within a roundel, and depict a lion struggling with the waves surrounded by flags, trophies of arms and a trumpeting angel. On the starboard side, the roundel contains the red castle of Aardenburg. Decorating the transom there is a charming painting of the prospect of a town, presumably Aardenburg, viewed from the sea. The carving centred in the upper counter just below the arched window may be a heraldic device referring to the owner of the yacht, but it is heavily coated with gesso and difficult to decipher. An imposing and graceful hexagonal lantern crowns the stern.

The emblematic references to the province of Zeeland and the town of Aardenburg suggest that the yacht may have been built in Aardenburg, either as a private vessel or a state yacht. The name of the yacht is unknown, but given the figures of hunters at the stern and quarters, it is likely the name referenced hunting in some way. Of interest, the original meaning of the Dutch word 'yacht' or 'jacht' (in the seventeenth century spelled 'jaght') was 'hunting' or 'hunt'. This type of vessel was called 'jacht' because of its ability to sail fast

A stairway just aft of the octagonal lantern leads down to the captain's cabin. An iron tiller travels in the narrow space between the maindeck and the floor of the pavilion before turning up to end with a wooden knob handle.

Baroque carvings incorporating cherubs, lions, dolphins and cornucopia adorn the quarter galleries and stern. The square tuck is typical of Dutch ships.

The pavilion is furnished with an octagonal table supported on a turned pedestal, just visible through the quarter gallery lights.

and hunt effectively. The hunted prey were often pirates who plied the shallow waters of the Low Countries. Eventually the vessels took on a more sporting role and were often raced by wealthy or noble individuals.

Origins of yachting

The first English yacht was the *Mary*, given to King Charles II by the Dutch East India Company upon his return to England in 1660. It was a 66ft vessel, finely decorated, provided with six 3pdr guns and leeboards in the Dutch style. The king was very fond of yachts, and he went on to commission a total of twenty-five more, a record number that has not been matched in well over 300 years. The King named his first yacht the *Mary*, in honour of his sister, the *Henrietta* for his mother and the *Saudadoes* for his Portuguese Queen. But others, namely the *Cleveland*, *Portsmouth* and *Fubbs*, were named for his mistresses. There is an interesting model of one of Charles II's yachts in the Musée National de la Marine, Paris, which includes painted interior decorations. These consist of painted paper panels depicting a bewigged figure with a strong resemblance to the King, frolicking with naked maidens. The King's yachts were not exclusively pleasure craft, as many of them saw service in time of war and accompanied the fleet in battle. The *Henrietta*, King Charles' favourite yacht, was sunk at the Battle of the Texel.

Right: The gaff rig is evident in this photo, with a square topsail, a loose-footed gaff and two head sails.

A DUTCH STATE YACHT c1690

This unfinished boxwood carving was intended to be the figurehead of a ship model. Queen Caroline is depicted seated with cherubs and allegorical figures of victory holding a crown above her head with tritons sounding trumpets below.

CHAPTER 28

Model figurehead for the *Royal Caroline* 1750

~ Acquisition ~

W E HAVE LEARNED MANY times over the years how important personal relationships struck with key dealers can be in forming a collection. The problem is that one does not always know ahead of time who the key dealers are. The only solution is to cast as wide a net as possible and let everyone you meet who may conceivably come in contact with the target of your obsession, know of your interest (and make sure he has your current address, phone number or email address). This object is a case in point.

Many years ago we had purchased a very fine painting of the *Britannia* by Isaac Sailmaker from an old master painting dealer with a shop in St James's, London. Since then, we always made it a habit to visit this shop when doing the rounds of other dealers and auction houses in and around Bond Street. These dealers almost never have anything of real interest to us, but it only takes one find to make it all worthwhile. Just such a discovery occurred on a recent occasion when Arnold entered this particular shop during a quick business trip to London and the owner exclaimed that he had been thinking of Arnold just two days before! He had attended a Saturday country auction and bought a little decorative carving for his personal collection. This dealer had a very good eye and was a collector of decorative wooden carvings, and when he saw this one he immediately recognised it as part of the decoration for a ship. He dated it to the reign of William and Mary at the end of the seventeenth century because it appeared to have the two sovereigns carved on either side. Arnold was, of course, very eager to see this miniature carving, and the dealer agreeably offered to bring it to his shop the next day whereupon Arnold bought it.

A unique feature of this carving is the wooden handle to which it is fitted. Unlike the carving, which is boxwood, the handle is soft pine and roughly carved. It is approximately round in section, has an odd slot cut into the lower half, and has a rusty nail driven through near the back end in order to repair an old split in the wood.

HISTORIC SHIP MODELS

The wooden handle fits neatly into a narrow slot in the sculpture that is designed to fit the stem of the ship.

The entire figurehead is carved from a single piece of boxwood, although parts, such as the arms and legs of the cherubs, might have been intended to be later additions.

Description

The carving, as it turned out, was a miniature boxwood model of the figurehead of a ship that we agreed was probably made during the reign of Queen Anne with her effigy repeated twice, once on each side. It was apparently intended for the bow of a dockyard model but was never finished and never mounted. Even though it is only 2in high, this figurehead includes no fewer than eight full figures. The queen, seated, is shown on both starboard and port sides holding an orb and sceptre in each hand with winged angels behind, trumpeting tritons below and cherubs holding a crown aloft topped by a strutting lion. Such complicated figurehead designs were typical of warships of the first few years of Queen Anne's reign and we consequently dated the carving to 1702–04 and recorded it as such in the first two editions of the book about our collection. However, in this case we were mistaken. Many readers may be aware that as decorative features such as carved and gilded decorations gradually diminished or disappeared altogether with the evolution of warship design, they nonetheless persisted on royal yachts. We can point to this trend as the admittedly weak excuse for our misdating this figurehead by forty-five years. Some years ago, we became aware of the painting by John Cleveley of the *Royal Caroline* yacht of 1749 at the Royal Museums Greenwich. A close look at the figurehead reveals a perfect match for our little model. We can now state with certainty that the female sovereign repeated to both port and starboard of the model represents George II's wife, Queen Caroline. The *Royal Caroline* yacht was built by Joshua Allin in 1749 at Deptford. In 1761, when George III dispatched the yacht to bring Princess Charlotte from Germany to England to become his future wife, the yacht was renamed *Royal Charlotte*.

THE MASTER CARVER AT WORK

Unbelievably, the figurehead is still to this day mounted on the temporary handle that the model maker had fashioned for holding it while it was carved. The handle is fitted into the slot carved into the figurehead to accept the stem of the ship and shows the rough chisel marks of its casual manufacture as well as the polished surfaces of repeated handling. Even though the little carving has suffered lost bits and pieces over the course of three centuries, this in no way diminishes the impact of such a remarkable survival. This glimpse into the creative process of carving the miniature decorations of an Admiralty ship model is entirely unique in our experience, and we doubt that another example survives. When looking at this little object, one can easily be transported to another time and place and can imagine the master carver himself at work, hunched over the tiny half-finished figures carefully wielding his miniature chisels and planes. The feeling is even more intense and tactile when one holds the carving by its well-worn handle. One wonders why it was never finished. Dissatisfaction? Design change? Infirmity? Death?

We have seventeen complete Admiralty Board models in our collection, decorated with an aggregate number of many hundreds of miniature human figures, mythical creatures and animals, but when it comes to appreciating the creative process itself, this little unfinished carving surpasses them all.

This is a detail from a painting of the *Royal Caroline* yacht by John Cleveley. The ship is identified by an inscription to the left of the canvas along with the artist's signature and the date 1750. Cleveley worked as a shipwright in the Royal Dockyard at Deptford and his depictions of ships are extremely accurate. The figurehead of the *Caroline* in the painting exactly matches our carving. Courtesy of Royal Museums Greenwich.

This maquette for the figurehead of the 1st rate *Queen Charlotte* is 8in tall. Carved c1785 at the Chatham dockyard, it has been in this bell jar since the nineteenth century and remains in excellent condition.

CHAPTER 29

Model of the *Queen Charlotte* figurehead c1784

~ Acquisition ~

THIS MODEL WAS HIDDEN from public view until a similar example came up for auction in London on 5 July 2005. That model, also a design for the *Queen Charlotte* figurehead, was sold by Bonhams and acquired by the Historic Dockyard, Chatham. Shortly afterward, the figurehead historian Richard Hunter was called to examine a lime wood carving kept in the basement of a Georgian town house in London by a descendant of Sir Andrew Snape Hamond. It was Richard Hunter's opinion that this magnificent carving was an earlier design for the *Queen Charlotte* figurehead, and 'must rank as the most important free-standing surviving model of a British Warship, and as such of International importance'. The owner subsequently decided to sell it, and I was fortunate enough to acquire it.

~ Provenance ~

The bell jar that encloses this model bears an old handwritten label, transferred from some other document, which reads:

> Model of the Figure Head of H.M.S. *Queen Charlotte* 110 guns Flag Ship of Adm. Earl Howe in the past Battle and Victory of the 1st June 1794. It being a full length of her majesty Queen Charlotte and a good likeness. Graham Eden Hamond Mids. & A.D.C. to the Captain on that day.

The model descended in the Hamond family, and Admiral of the Fleet Sir Graham Eden Hamond Bt (1779–1862) was indeed the fifteen-year-old aide-de-camp to the captain of the *Queen Charlotte* at the battle of the Glorious First of June. The model is clearly a design intended for the *Queen Charlotte* of 1790, which was a 100-gun ship. The label confuses her with the second *Queen Charlotte*, launched in 1810 and carrying 110 guns. The label was clearly

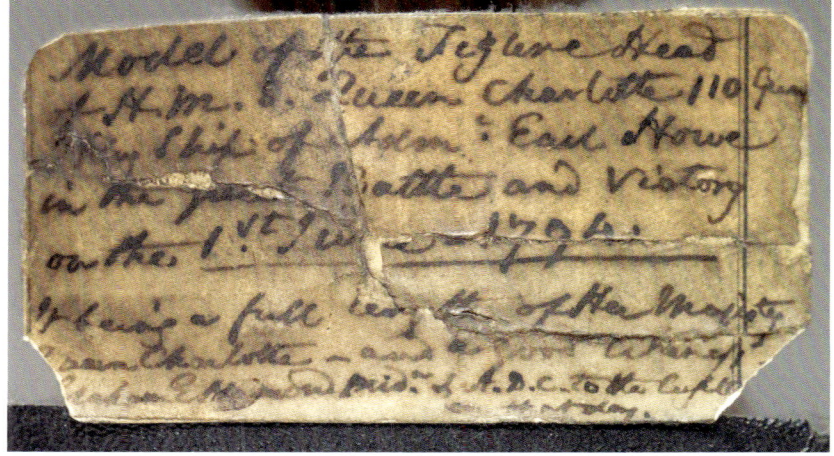

This old label is now attached to the Victorian bell jar that protects the model. The author correctly states that Admiral of the Fleet Sir Graham Eden Hamond was aide-de-camp to the captain of the *Queen Charlotte* (his uncle) at the Battle of the Glorious First of June, when he was just fifteen years old. However, he had inherited the model from his father, Sir Andrew Snape Hamond, who was Commander-in-chief in the River Medway and at the Nore when the ship was being built at Chatham.

written after 1810, at a time when the figurehead belonged to Graham Hamond many years after it was created. The model would have been inherited by Graham, as it was his father, Sir Andrew Snape Hamond Bt, who was in a position to acquire the maquette after it had served its purpose. It was also Andrew Hamond who was responsible for gaining his son's appointment aboard the *Queen Charlotte*, because it was Andrew Hamond's nephew, Sir Andrew Snape Douglas, who was commanding Earl Howe's flagship the *Queen Charlotte* on that celebrated day.

Andrew Hamond was born in 1738 and entered the naval service in 1753. During the Seven Years' War he saw action under Earl Howe and the Duke of York and attained the rank of post-captain on 7 December 1770. He served on board the 90-gun *Barfleur* before taking command of the *Arethusa* frigate. He saw hot action during the War of American Independence and played a key role in the taking and holding of Philadelphia in 1777. He was Captain of the Fleet at

This is a portion of the original maquette for the figurehead of the *Queen Charlotte*, which was launched at Chatham in 1790. The model dates from c1785 and is carved out of lime wood. It became the property of the Comptroller of the Navy at the time, Sir Andrew Hamond, Bt, and descended in his family until 2017. It measures 8in high overall.

There are two different symbolic figures flanking the Queen and four different anthropomorphic Virtues at her feet, yet the designer of this model arranged them all with near perfect symmetry.

the reduction of Charlestown, and late in 1780 he was appointed Lieutenant Governor and Commander-in-chief of Nova Scotia. He returned to England in 1783 and was created a Baronet of Great Britain for his distinguished services in North America. It is the next phase of his career that is pertinent to us, because in 1785 he became Commander-in-Chief in the River Medway and at the Nore, headquartered at Chatham. He was in charge of Chatham dockyard while the *Queen Charlotte* was being built there, and this is likely when he acquired the figurehead maquette. This was not his only opportunity to do so, however. In 1793 he became a Commissioner of the Navy Board, and Deputy Comptroller of the Navy in February 1794. In August of that year he was appointed Comptroller and held that office until his retirement in 1806. It is likely a coincidence that his nephew and son should subsequently see action on the ship constructed under his watch.

From Baronet Andrew Snape Hamond the model descended in the family for over 300 years until we had the opportunity to acquire it.

Description

CONDITION

The model currently sits in a glass-domed case that has protected it since the nineteenth century. Prior to that it must have been kept

MODEL OF THE QUEEN CHARLOTTE FIGUREHEAD c1784

The principal figure on the starboard side represents winged Victory, offering a laurel, holding a torch and with a shield at her feet.

The principal figure flanking the Queen on the larboard side represents Plenty with a cornucopia in her left hand. Above her head is a winged cherub blowing a clarion.

out of harm's way in some other container, because it is in a wonderful state of preservation. This diminutive model measures only 8in high and is incredibly delicate, yet is complete and appears to be almost entirely original. The right hand of the figure representing Tolerance on the starboard side of the carving appears to be a replacement, as does the top of Queen Charlotte's sceptre.

CONSTRUCTION

The model has been masterfully assembled out of several pieces of lime wood, unvarnished and held together with glue and wooden pegs. It features a central carving of the Queen in her Coronation robes holding an orb and sceptre, surrounded by eight allegorical figures. Each element is delicately and masterfully carved and posed to create a wonderfully balanced and aesthetic ensemble. To keep the structure light there had to be space between its many elements, and the overall achievement is a testament to the skill and talent of the artist.

Although we cannot say with certainty who carved this model, there are two chief possibilities. William Savage was master carver at Chatham from 1765 until 1783, and he may have produced this carving early in the design process. The other candidate is his successor as master carver, George Williams, who worked at Chatham from 1784 until 1834.

Literature

The following reference includes a description of this model: *Royal Naval Exhibition* (London, W P Griffith & Sons Ltd, 1891), p. 321.

Exhibitions

Royal Naval Exhibition, Chelsea, 1891, No. 3219, Lent by Sir Graham Eden Hammond-Grame, Bart.

Historical Perspective

DESIGN OF THE FIGUREHEAD FOR THE *QUEEN CHARLOTTE* OF 1790

This model celebrates Sophia Charlotte of Mecklenburg-Strelitz, the wife of King George III and the Queen of England from 1761 until 1818. She was forty-eight when the *Queen Charlotte* was ordered in 1782, and fifty-six by the time she was launched at Chatham in 1790. Numerous portraits of this popular queen exist, and comparison with the effigy of her on this model confirms that the likeness is good, as the old label claims.

First-rate ships of the line were viewed as ultimate ambassadors and symbols of national pride and prestige, and their figureheads were contrived to impress and intimidate. In his classic work on *Old Ship Figureheads and Sterns*, L G Carr Laughton records that the figurehead of the 1790 *Queen Charlotte* was regarded by contemporaries 'as being the most handsome that was ever put into a ship'. It may not surprise us, then, that three contemporary models of this complex figurehead exist, along with drawings and even the written order for the final design. Two of the models are free-standing: ours and the one at Chatham Dockyard; and the other is on a full-hull Navy Board model of the ship in the collections of the Royal Museums, Greenwich. These models differ in interesting respects, and examination of the variations can help place them in sequence and trace the evolution of this noteworthy figurehead's design.

Our investigation of this evolution will begin with the end result, viz. the official instructions to the carvers of the actual figurehead, issued when the design was finalised. It reads as follows:

Beneath the Queen on the larboard side is a representation of Prudence with her mirror, and behind her is a personification of Fortitude with her left hand on Samson's column.

> In the head is Her Majesty in her robes with orb and sceptre in her hands, standing erect under a canopy with two doves thereon, which is supported by two boys, the emblems of peace, one holding a dove, the other a palm branch; under which on the starboard side is Britannia sitting on a Lion and presenting a laurel; on the larboard side is Plenty sitting on a sea-horse offering the produce of the sea and land; on the starboard trail board Justice and Prudence with their emblems; on the larboard trail board are two boys, Hope and Fortitude, with their emblems.[1]

The first, 'oldest' iteration is our maquette. It features eight allegorical figures flanking the Queen, but no animals. It differs the most from the final specification, which supports its ranking as the earliest of the proposals. Many of the final elements appear in this

The figures beneath Victory on the starboard side represent two of the Cardinal Virtues: Temperance holding her bridle and reins; and behind her sits Justice with her emblematic sword.

version, beginning with Queen Charlotte herself, depicted standing under a canopy in her coronation robes and holding the orb and sceptre. She is announced by a pair of winged putti blowing trumpets. On the starboard side she is flanked by a winged Victory with her shield and offering a laurel. On the port side she is flanked by Plenty, offering the produce of the sea and land. Beneath these representations there are allegorical figures representing four of the cardinal Virtues. On the starboard side there is Temperance with her bridle and reins, and behind her rests Justice with her sword. On the larboard side there is a representation of Prudence with her mirror, and behind her rests Fortitude with Samson's column. Behind and around these figures there are very delicate foliate and scroll carvings.

The second model in sequence is the one at Chatham Dockyard, which descended in the family of Sir John Henslow, Chief Surveyor to the Navy from 1774 to 1806. It is also made out of lime wood, and is a little larger than ours, but about half of the original elements are missing. Nevertheless, we can see that some changes to the design have been made. The canopy is gone, so we cannot comment on whether there were birds upon it or putti beneath. On the starboard side the principal figure has also been lost, so we cannot determine whether Victory or Britannia once flanked the queen. Beneath this missing emblem a lion has appeared, with two Virtues beneath and alongside him, but the distinguishing accoutrements they once held are sadly missing so that their identity also cannot be determined. On the larboard side the principal figure is also missing, but the sea horse has made his appearance beneath the missing flanking figure, alongside what were once two more Virtues. Unfortunately, one of these is altogether lost, and the remaining personification has lost her right hand along with the identifying symbol it would have held. Thus, there are still four Virtues present, but we cannot say whether they are the same ones that appear on the earlier version. The overall arrangement has been preserved, however, and the carving would have originally held eight figures surrounding the Queen.

The third and final model of this figurehead to consider is the small (¼in = 1ft) boxwood one on the Navy Board model of the ship at Greenwich. This model is closest to the official instructions prepared for the carvers and shows that additional changes were made to achieve the final perfected form. The main difference from the instructions is the absence of the doves and the emblems of Peace (which do not appear on our model either), but compared to the earlier models, there are other significant discrepancies. The trumpeting putti are gone, and the four Virtues have been removed from the figurehead and shifted instead on to the trailboard. On the trailboard, the starboard side still features Justice, but she is now accompanied by Prudence, who has moved from the larboard side. The larboard side has retained Fortitude, but Temperance has been replaced by Hope. It is interesting to see how the shifting of these four elements off of the figurehead itself serves to reduce the complexity and alter the profile of the carving. Clearly these changes represent improvements from the functional perspective, but whether they represent an aesthetic improvement is open to debate.

There is another free-standing, complex and beautiful model of a figurehead, which is in the collection of the Royal Museums, Greenwich. It represents the figurehead of another 100-gun ship, the *Victory* of 1765, and is made out of boxwood to a scale of approximately 1:24. It has been claimed that this is a design for the

actual figurehead, but this cannot be true. Integral to the carving is a shield with the Union flag, and it clearly features the St Patrick saltire, dating the piece to no earlier than 1801. The *Victory* underwent a great repair from 1800–03, during which this original figurehead was replaced with a much simpler design. The model was likely made around that time to document the original carver's achievement. It is not a maquette, as it post-dates the carving of the original figurehead. Our model of the *Queen Charlotte* figurehead is therefore likely the earliest known maquette for an English ship figurehead.

HISTORY OF THE
QUEEN CHARLOTTE OF 1790

It took eight years to design and construct the 100-gun 1st-rate ship of the line *Queen Charlotte*. She was launched at Chatham Dockyard on 15 May 1790. Her gun decks carried 32pdr cannon,

Larboard side of the *Queen Charlotte* figurehead. It is remarkable that three models exist that trace the design stages for the *Queen Charlotte* figurehead. These different-sized models have been digitally equalised to facilitate comparison. On the left is our model c1785. It features the Queen flanked by Plenty and a trumpeting angel, and standing above two cardinal Virtues, Prudence and Fortitude, all beneath a parasol canopy. The middle image shows a model built to a larger scale, which is closer to the final configuration. It is missing many elements, but a sea horse has appeared alongside the Virtues. The figure on the right is from an Admiralty Board model of the ship as built in 1790 and it shows that Plenty is now riding on the sea horse, and with regard to the Virtues, Temperance has been replaced by Hope and both have been relegated to the trailboard. All of this is surmounted by a more restrained canopy. Centre image courtesy of Bonhams, and right image courtesy of Royal Museums Greenwich.

MODEL OF THE QUEEN CHARLOTTE FIGUREHEAD c1784

Starboard side of the *Queen Charlotte* figurehead. These different-sized models have been digitally equalised to facilitate comparison. These three models of the *Queen Charlotte* figurehead offer unique insight into the evolution of this masterful design. Our model, on the left, is the earliest iteration. The Queen in her coronation robes is announced by a trumpeting winged cherub, and she is flanked by a symbol of Victory. Below these figures are representations of the Virtues Temperance and Justice. In the centre is a photograph of a larger-scale model showing an intermediate stage in the design. While many of the figures are sadly lost, a lion has been fitted beneath the missing flanking figure. On the right is an image of the figurehead as it appears on an Admiralty Board model of the completed ship. Here we see the trumpeting putti are gone, and Temperance has been replaced by Prudence, while both Virtues have been shifted off the figurehead and are carved in relief on the trailboard. Centre image courtesy of Bonhams, and right image courtesy of Royal Museums Greenwich.

32pdr carronades, as well as 24pdrs and 12pdrs. She was manned by a crew of almost 900 and was the largest British built ship afloat. As the flagship of Admiral Lord Howe, she saw action during the French Revolutionary War, famously leading the fleet at the Battle of the Glorious First of June, 1794. On 29 May, captained by Sir Andrew Snape Douglas and with Graham Eden Hamond on board, the *Queen Charlotte* led the British fleet against the French under Rear Admiral Villaret-Joyeuse and was the first to break the French line that afternoon. This action proved indecisive, and the battle resumed in earnest on the morning of 1 June. This time Lord Howe took the *Queen Charlotte* through the French line astern of Villaret-Joyeuse in the *Montagne* and delivered a raking broadside. Subsequently engaged by both *Jacobin* and *Montagne*, *Queen Charlotte* lost her topmast, but the French flagship suffered 300 dead or wounded. When the smoke cleared, Lord Howe had gained a great victory with seven French ships sunk or captured. After the victorious fleet had returned to Spithead, the King and royal family boarded the *Queen Charlotte* to personally express their gratitude to Lord Howe.

The *Queen Charlotte* next distinguished herself off Isle Groix on 23 June 1795. Lord Howe being ill, the fleet command fell to Lord Bridport in the *Royal George*, while the *Queen Charlotte* was again captained by Sir Andrew Snape Douglas. Villaret once again led the French, and in the action that ensued the *Queen Charlotte* 'distinguished herself above all other ships that day'. The British were victorious, but a chance to annihilate the French was squandered by Bridport, much to his disgrace.

On 17 March 1800, disaster struck the *Queen Charlotte*. She was about 12 miles out of Livorno when a fire broke out a little before 6am. A live match was kept in a tub under the half-deck for firing signal guns. Hay left lying nearby caught fire and the flames soon raged out of control. Eleven persons were on shore and escaped the conflagration, including Lord Keith, the Vice Admiral and flag officer at the time. Several American vessels lying at anchor off Leghorn offered assistance, but the heat of the fire caused many of the ships loaded guns to fire, killing several of the brave rescuers. Attempts to extinguish the flames were futile, and about 11am the ship blew up. A total of 156 British seamen were saved, but 673 perished, and the great ship with its beautiful figurehead was gone.

Right: This drawing by J Perriman is a posthumous depiction of the *Queen Charlotte* of 1790 and corresponds exactly to the model of the ship in the Royal Museums Greenwich collection. It is noteworthy that the ship is depicted as a model.

Below: There was a vogue for complex symbolic figureheads on 1st rates in the late eighteenth century, and this design for *Britannia* is another example.

MODEL OF THE QUEEN CHARLOTTE FIGUREHEAD c1784

J. Perriman 1807. Queen Charlotte 110 Guns.

Composite view of the four sides of the model. It is contained in an oak box with sliding lid, also made of wood removed from *Victory* when she underwent repairs after the Battle of Trafalgar.

CHAPTER 30
Model of the foremast of the *Victory* of 1765 with damage sustained at Trafalgar

Acquisition

SHIPS HAVE BEEN SUNK, wrecked and battered along the English coast and in her waterways ever since the first Britons ventured out in their canoes. In recent centuries their descendants, ever resourceful, have found ways to profit from the misfortunes of their hapless brethren. If the ship that fell victim to some calamity was famous enough, then bits of her wreckage could be marketable, especially if recycled into something useful, decorative or interesting. This practice of selling naval relics reached its apogee in the nineteenth century when pieces of famous old ships were fashioned into cannon, books, models, furniture, etc. When the 1st-rate *Royal George* sank in the Solent while undergoing routine maintenance on 29 August 1782, more than 800 lives were lost. The notoriety of this catastrophic event helped create a market for a vast number of souvenirs made from bits of her wreckage; so many that it has been said two *Royal Georges* could be constructed from the amount of timber in these tokens.

Another of the most famous warships of the age of sail is the *Victory* of 1765, known as 'Nelson's *Victory*'. She was his flagship at the Battle of Trafalgar, where both he and the ship became the target of withering fire from the French. Nelson succumbed to his wounds, but his flagship survived to be towed back to England. Much of the ship needed to be replaced, and this

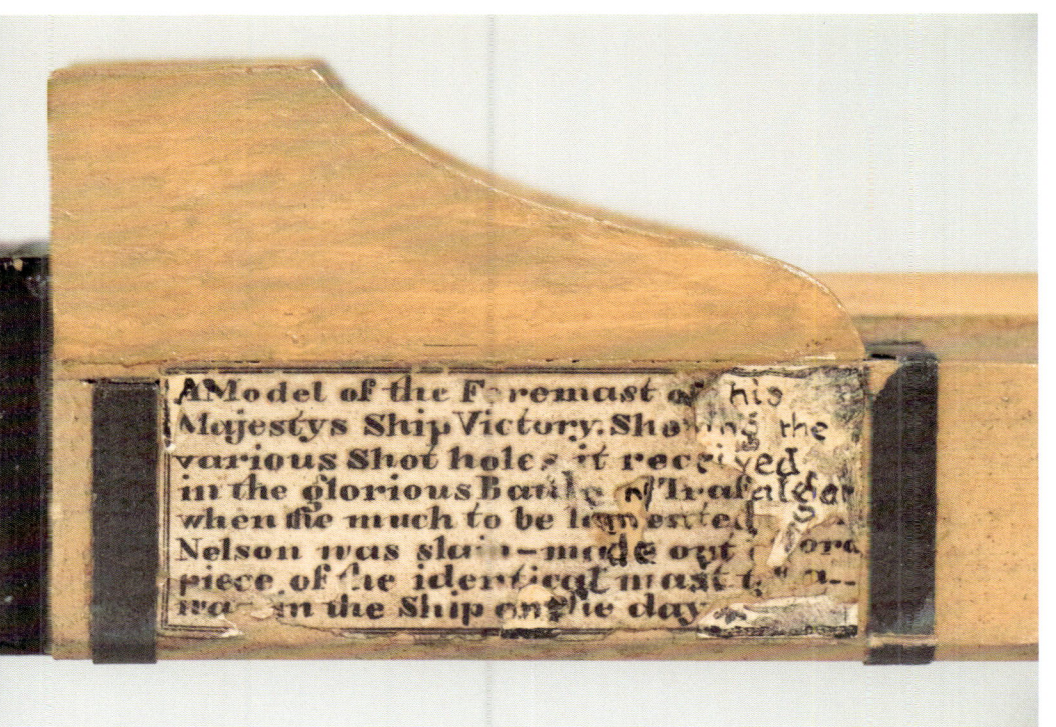

The original handwritten label attached to the hounds of the foremast describing the model.

There is a scale indicator at the base of the mast head, and there are labels indicating the point at which the mast passes through each of the decks.

Detail at the lower end of the cheeks showing gouges and impact scars from cannon fire damaging the cheeks and front fish.

relic is the result of an enterprising model maker who got hold of some of the discarded timber.

The model accurately represents the foremast of the *Victory* with all the damage she sustained at the Battle of Trafalgar faithfully recorded. It is made out of a piece of the mast removed when she was towed to Chatham for repairs and is enclosed in a box made of oak also removed from the ship at that time. It is 4ft 5in long and a handwritten original note attached to the mast states that it is:

> A Model of the Foremast of his Majestys [sic] Ship *Victory*, Showing the various Shot holes it received in the glorious Battle of Trafalgar when the much to be lamented Lord Nelson was slain – made out of a piece of the identical mast that was in the Ship on the day of Battle.

Of the many old ship relics we have seen, we regard this historical artefact as the most interesting and important. It is not unique, however. Two nearly identical examples have long been in the collections of the Royal Museums, Greenwich (SLR 2468, SLR2485). We were surprised when another one turned up at a country auction in England early in the twenty-first century and were delighted to add this one (the subject of this chapter) to our collection.

Provenance

This model was in the Royal United Services Museum by 1914 and is described in the official catalogue as 'Model of the foremast of the Victory made from wood of the ship for Vice Admiral John Drake, who was present as midshipman and master's mate on board the "Defiance" both in Sir Robert Calder's action (22nd July, 1805), and at Trafalgar … The model having passed to his son, came subsequently into the possession of the donor as the next representative.'[1]

It passed into the collection at Kilcoy Castle, Inverness, family seat of the Mackenzies, and we bought it when they put it up for auction. It bears a label that states it was 'originally exhibited at the Royal United Services Museum, Whitehall'.

Literature

The following reference includes a description of this model:
Leetham, Arthur, *Official Catalogue of the United Service Museum* (Southwark: J J Keliher & Co., 4th edition, 1914), p. 88d.

Exhibitions

Royal United Services Museum, Whitehall, prior to 1914 until 1964.

MODEL OF THE FOREMAST OF THE *VICTORY* OF 1765 WITH DAMAGE SUSTAINED AT TRAFALGAR

Description

CONDITION

The mast remains in its original bespoke case, which has protected it from many of the ravages of time. It has lost some of the paper mast bands that represent the wooldings that encircled the mast at intervals from tenon to hounds. It is otherwise in excellent condition, as is the box.

CONSTRUCTION

Scale: 1/24 Length of Mast: 4ft 5in

The box, with its sliding lid, is fashioned from oak allegedly removed from the *Victory*. The mast itself is made of fir taken from the foremast in 1805 after the Battle of Trafalgar. Great care has been taken to accurately reproduce the damage from shot and ball that this mast sustained on 21 October 1805.

The model is of the fore lower mast belonging to the 1st-rate *Victory* of 1765. It is quite skilfully made and is complete from the step of the heel tenon to the top of the head. There are labels affixed to the front that mark where the gun deck, middle deck, upper deck and forecastle would intersect the mast. Rope wooldings were replaced on the *Victory* by iron bands in 1803, and these are represented by bands of paper painted black. Structural features include the rubbing paunch fitted to the front of the mast extending from the stop of the hounds to nearly the forecastle deck; both port and starboard cheeks; the hounds with bibbs attached; and the rectangular tapering head of the mast. The head is painted black, but the rest of the mast is simply varnished.

The most important aspect of this model, however, is the way damage to the mast has been painstakingly rendered. Small features have been faithfully carved to document the condition of the mast at the end of the day on 21 October 1805. Ragged shot holes, splintered cheeks, jagged scars left by glancing cannon balls, are all recorded in sobering detail. The model effectively conveys the dreadful destruction wrought on both sides during that momentous battle.

It is not known for certain who made this model, but the builder obviously had access to the actual mast since it was both the source of his materials and the prototype for its representation. It is therefore likely to have been built by a model maker at Chatham and can be considered a dockyard model.

Another view of the damage sustained near the middle of the foremast, revealing how close the French gunners came to bringing down the mast.

Carefully rendered battle scars from perforating and penetrating shot strikes near the hounds.

Historical Perspective

VICTORY AT THE BATTLE OF TRAFALGAR

On 13 December 1758, the Board of Admiralty ordered the construction of a 100-gun ship, which was to become the most successful 1st-rate man-of-war ever built. She was the fifth Royal Navy vessel to be named *Victory*, and she is still in commission today. Designed by Thomas Slade, her keel was laid down at Chatham Dockyard on 2 July 1759, but she wasn't ready for launching until 7 May 1765. She proved to be an excellent sailer and served as the flagship of a succession of illustrious admirals, including Richard Kempenfelt, Lord Howe, Lord Hood, Lord St Vincent and Sir James Saumarez. Her most famous commander, however, was of course, Horatio Nelson. But Nelson first took

Painting of the *Victory* at sea c1770. Her altered appearance at the Battle of Trafalgar reflects the results of a major repair lasting from 1800 to 1803.

command of her as Commander-in-chief of the Mediterranean fleet in 1803 when the ship was thirty-eight years old, well beyond the expected lifespan of a wooden warship.

Victory had survived so long thanks to a series of repairs and rebuildings, most notably a major reconstruction carried out in 1801–03. It was in this rebuild that her fir masts were replaced with the ones she carried into battle at Trafalgar, a piece of which ultimately turned into this model. It was on 21 October 1805, off Cape Trafalgar along the south-west coast of Spain, that *Victory* sailed into battle and into the annals of history.

The battle scars suffered by the foremast testify to the pummeling the *Victory* received that day.

Admirals Nelson in the *Victory* and Collingwood in the *Royal Sovereign* each led a column of British vessels across the French line, and so bore the brunt of the fighting. Nelson was mortally wounded and the *Victory* nearly disabled. The damage to the ship was listed by Midshipman R F Roberts after the battle:

The hull is much damaged with shot in a number of different places, particularly in the wales, strings, and spurketing, and some between wind and water. Several beams, knees, and riders, shot through and broke; the starboard cathead shot away; the rails and timbers of the head and stem cut by shot; several of the ports damaged, and port timbers cut off; the channels and chainplates damaged by shot, and the falling of the Mizzen mast; the principal part of the bulkheads, halfports, and portsashes thrown overboard in clearing ship for action.

The mizzen mast shot away about nine feet above the deck; the mainmast shot through and sprung; the main yard gone; the main topmast and cap shot in different places and reefed; the main topsail yard shot away; the foremast shot through in a number of different places, and is at present supported by a topmast and a part of the topsail and crossjack yards; the fore yard shot away; the bowsprit, jibboom and cap shot, and the spritsail and spritsail topsail yards, and flying jibboom gone; the fore and main tops damaged; the whole of the spare topmast yards, handmast, and fishes shot in different places, and converted into jury geer.

The ship in bad weather taking in 12 inches of water an hour.[2]

Examples of some of the souvenirs of the *Royal George* made from recovered wreckage.

MODEL OF THE FOREMAST OF THE *VICTORY* OF 1765 WITH DAMAGE SUSTAINED AT TRAFALGAR

The *Victory* was not in sailing condition, and was towed to Gibraltar for temporary repairs, thence to Portsmouth, and eventually to Chatham for the second major refit of her career. It must have been at Chatham where a skilled model maker, recognising the historic opportunity, decided to make a set of miniature replicas out of the foremast timbers. The *Victory* herself, the most venerable of memorials, sits in dry dock in Portsmouth, but much of her original structure has succumbed to the ravages of time and is gone. The little foremast model, however, preserves a piece of her that has remained, and will remain, essentially unchanged for centuries.

This detail from a painting by Captain William Elliott shows shipping in the Solent with the masts of the sunken *Royal George* in the right foreground. She was accidentally sunk while heeled over for repairs on 29 August 1782, with the loss of over 800 lives. Her remains were a hazard to shipping for many years until she was finally broken up in 1840.

This is the earliest known painting of an Admiralty Board ship model and depicts the 1st rate *Royal William*, built at Portsmouth and launched in 1719. She was never sent to sea and remained laid up in Portsmouth harbour. She served as a tourist attraction until 1756, when she was cut down to an 84-gun 2nd rate.

CHAPTER 31
Ship models in perspective painted on panels

During the last thirty years or so, we have come across a number of eighteenth-century portraits of naval ships depicted as models. Occasionally these have been drawings on paper or vellum, but the most common form is an oil painting on panel, canvas or copper plate. We have acquired six paintings of ship models on panel and two on canvas, including the oldest known example, a broadside view of the *Royal William* of 1719. This is a unique representation, painted on paper and laid down on wood, and it is dated in a cartouche 1729. Unfortunately, it bears neither a signature nor a monogram, and the artist is unknown.

The two portraits on canvas in our collection are among the oldest *pair* of perspective ship model depictions known and represent the *Victory* of 1738. Armed with over 100 brass guns, she was the largest ship afloat in the world at the time of her launching, but her career was tragically cut short during a storm in October 1744 when she struck an outcrop of rocks near the island of Guernsey and sank. John Entick offers the following comment on this tragic event:

> A Misfortune sensibly felt by the Public, and greatly deplored by the Private, because the *Victory* carried 110 Brass Guns; and, for her fine Dimensions, and rich Ornaments of Painting and Gilding, she was esteemed the most beautiful Ship in the Royal Navy; not only the largest Ship in the British Fleet, and the finest Set of Guns, Masts, Rigging, and Yards, went to the Bottom.[1]

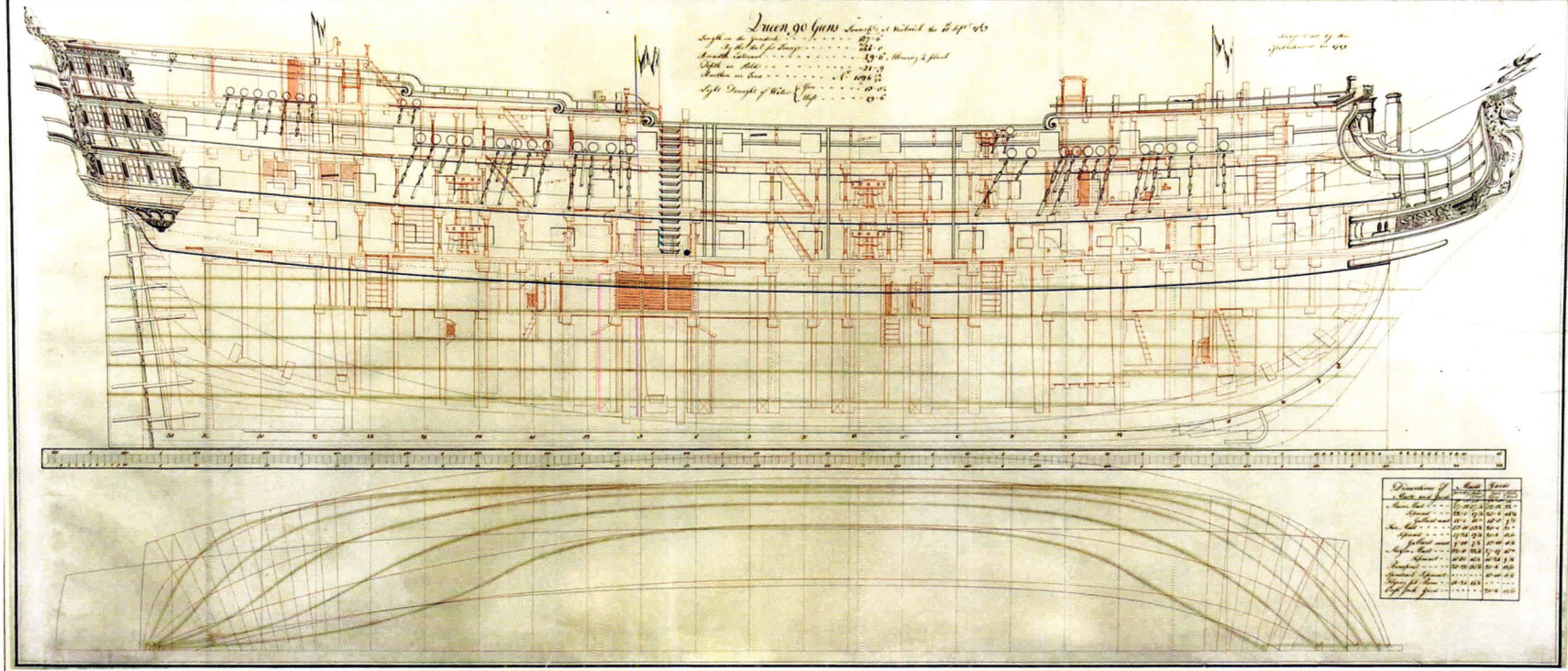

This is a draft of the *Queen*, a 2nd rate built at Woolwich dockyard and launched on 18 September 1769. It may be the draft shown to King George III in 1773 as a prototype for a series of representations of men-of-war. It was rejected in favour of ship model portraits.

This pair of paintings on canvas depict a model of the *Victory* of 1738. This 1st rate, the largest wooden ship afloat at the time, is known as 'Balchen's' *Victory* because Sir John Balchen was the admiral in command when she sank. This tragic loss occurred in a gale on the night of 10 October 1744, while she was homeward bound in the English Channel. She vanished with 1,100 souls on board, and no trace of her was seen for the next 264 years, until she was located on the sea floor in May 2008. The paintings went to Daniel Finch, 3rd Earl of Nottingham and 8th Earl of Winchelsea, who was First Lord of the Admiralty in 1742–44. We purchased them from his descendants, in untouched original condition, having remained in their original stretchers and frames for 265 years.

It was believed she went down near the Casquets, but the wreck has recently been found around 60km away. The Admiralty conducted an inquiry immediately after the disaster, and these paintings may have been produced in connection with it. Two nearly identical pairs of paintings of the *Victory* were created; one pair went to the royal collections at Windsor Castle and the other to the First Lord of the Admiralty at the time, Daniel Finch, 3rd Earl of Nottingham and 8th Earl of Winchelsea. This latter pair descended in the Earl's family until we purchased them in 2007. They resemble our older *Royal William* painting insofar as they are set in a domestic interior and are framed on one side by a drape. The *Victory* paintings made for the King are now at the Science Museum, London, and show the model against a stark black background. It is interesting that both the *Royal William* and the *Victory* were built at Portsmouth.

This reinforces our belief that these paintings were all the work of the same anonymous artist. It is also worth noting that there is a model of this *Victory* at the NMM, built for the Admiralty Board Room c1744, and it is displayed on the same distinctive cradles as depicted on the *Victory* paintings. There is another dockyard model of this ship at Cawdor Castle near Inverness in Scotland, and it shares other features with these paintings, including the unusual position of the entry ports.

Nine of the paintings in our collection are related to a project undertaken later in the eighteenth century at the behest of King George III. In 1773, the King requested that the 4th Earl of Sandwich, who was First Lord of the Admiralty, prepare plans of one example of each class of naval vessel for the King's use. Not sure of what format the King might prefer, in August of that year,

This bow view of the *Royal George* was painted by Joseph Binmer and John Marshall and is dated 1779. The companion stern view is at the Royal Museums Greenwich, England. The *Royal George* was a 1st rate launched at Woolwich dockyard in 1756. She was admiral Sir Edward Hawke's flagship at the Battle of Quiberon Bay on 20 November 1759, and was with Admiral Sir George Rodney's squadron off the coast of Spain on 16 January 1780, when six Spanish ships were taken after a sharp action during which the Spanish 70-gun *Santo Domingo* was blown up. While preparing for the relief of Gibraltar in 1782, the *Royal George* was heeled over for repairs at Spithead when she capsized and sank at her anchors on 29 August with the loss of over 800 lives.

the Earl sent King George both a draft of the *Queen*, a 90-gun ship built in 1769, and two perspective paintings of a model of the *Berwick*, a 70-gun ship built in 1743. The King chose the perspective paintings and a list of twelve ships to be represented in this manner was submitted by Sir John Williams, surveyor of the Navy, in January 1774. The work was performed by a team consisting of Joseph Williams, who drew the stern views; John Binmer, who drew the bow views; and Joseph Marshall, who rendered them in oils. Navy Board records indicate that Binmer was an assistant to the surveyor, and that Marshall was paid £21 in 1774 'for painting two views of the *Barfleur*', which is model number 5 in the series.[2] The project was completed on 25 August 1775 and ultimately portrayed fifteen vessels. Queen Victoria in 1864 presented the perspective portraits of eleven of them to the Science Museum, South Kensington, where they remain today.

In subsequent years, this same team produced additional pairs of perspective paintings, some depicting ships drawn from the original list and others depicting new vessels. A few were evidently produced in 1779, as a number have survived bearing that date. Two are now in the Royal Museums, Greenwich, and seven in our collection.

It is by remarkable good fortune that some years ago a London dealer offered us a draft of an eighteenth-century naval ship. It turned out to be the 1769 draft of the *Queen*, 2nd rate of 90 guns, and is undoubtedly the 'prototype' proffered to King George in 1773. It is part of our collection now.

These paintings on panel are by Joseph Binmer. The 74-gun 3rd rate *Hector* was launched at Deptford in 1771. She fought in the van at the Battle of Ushant on 27 July 1778, and at the Battle of the Saints on 12 April 1782, where she was hotly engaged and dismasted, but managed to avoid capture. She ended her days as a prison ship in Plymouth and was broken up in 1816.

HISTORIC SHIP MODELS

Bow and stern views of an unidentified 5th rate c1770. Attributed to Joseph Marshall.

PERSPECTIVE PAINTINGS OF EIGHTEENTH-CENTURY MEN-OF-WAR IN THE KRIEGSTEIN COLLECTION

Royal William, 1st rate of 1719

Broadside view, Oil on panel, 62in by 29in. Bears the following information in a cartouche:

> *Royal William*
> Length for tonnage…140ft 7in
> Length for ye gun-deck…174ft 0in
> Breadth extreme…50ft 0in
> Draught of water afore…21ft 6in
> Draught of water abaft…22ft 10in
> Burthen, 1869 40/94
> MDCCXXIX

This painting was once owned by the author, Walter Wood, who described it in his book *The Battleship*, published in 1913. It was acquired from Mr Wood by Colonel Henry H Rogers in England sometime before 1932. We obtained it from his heirs in a trade in 1999.

Starboard-quarter bow view of the frigate *Enterprise*, carrying 28 guns and 200 men. The companion stern view is in the collection of the Royal Museums Greenwich. Both were painted by J Marshall in 1777 and are on panel. When we bought this painting it was black with dirt and had been flipped over and built into a desk as a writing surface!

~ Victory, 1st rate of 1738 ~

Bow and stern views, Oil on canvas, each 26in by 24in. Original owner was Daniel Finch, 3rd Earl of Nottingham and 8th Earl of Winchelsea, who was First Lord of the Admiralty in 1742–44. We purchased the paintings from his descendants.

~ Royal George, 1st rate of 1756 ~

Bow view, Oil on panel, 29in by 45in. Dated 1779 and signed in the lower left corner 'J. Binmer D.' and in the lower right corner, 'J. Marshall P.' The painting was consigned to Christie's auction house by an antique dealer and sold in March 1942 to M Witt. It was sold at auction again in 1995, when we bought it. The matching stern view was presented to the Royal Naval Museum, Greenwich, in 1897 by a Mr W G Porter of Croydon and remains at the NMM.

Another enigmatic painting of a frigate by Joseph Marshall that relates to the Williams/Binmer series commissioned by King George III.

~ Hector, 3d rate of 1774 ~

Bow and stern views, Oil on panel, each 22in by 30in. Stern view is signed in the lower right corner 'J. Binmer *Del et Pinx't.*' Bought from the Rutland Gallery in London in 1987.

~ Unidentified Frigate c1770 ~

Bow and stern views. Oil on panel, each 24in by 17½in. Unsigned. Purchased from the Rutland Gallery in 1987.

~ Unidentified Frigate c1770 ~

Stern view only. Oil on panel, 24in by 17½in. Purchased at auction in London.

~ Frigate *Enterprise* 1776 ~

Bow view only. Oil on panel, 22in by 16½in. Purchased at auction as part of a desk.

~ *Valiant* 74 guns of 1759 ~

Bow view. Oil on panel. 10½in by 11in. Purchased at auction in Paris.

This perspective painting of a model shows the 74-gun *Valiant* of 1759, and has been attributed to J Perriman, who designed a pair of engravings of the same ship.

Literature

The following references include photographs and descriptions of these paintings:

Walker, Grant H, *The Rogers Collection of Dockyard Models*, Vol. 1 (Florence, OR: SeaWatch Books LLC, 2015). p. 40.

Winfield, Rif, *First Rate* (Barnsley, Seaforth Publishing, 2010), p. 46, 54, 57, 67, 86, 134.

Witt, M, 'A Picture of the *Royal George* of 1756', *The Mariner's Mirror*, 39, 1 (February 1953), pp. 58–60, plate 5.

Wood, Walter, *The Battleship* [for the *Royal William*] (New York: E P Dutton & Co, 1913), pp. 67–68.

References

Clowes, Laird, *Sailing Ships Their History and Development as Illustrated by the Collection of Ship-Models in the Science Museum, Part II* (London: His Majesty's Stationery Office, 1936), pp. 46–47.

DRAWINGS OF SHIP MODELS FROM THE EIGHTEENTH CENTURY IN THE KRIEGSTEIN COLLECTION

Anonymous artists depicted ship models on paper and vellum, and we have acquired a number of them over the years.

Below left: A draught of the Dutch warship *Boreas* launched for the Admiralty of Amsterdam in 1768. It was drawn by Willem Lodewijk van Genth in 1767 and shows the ship as a model resting on a plinth.

Below right: This small pen and ink drawing shows the 1st rate *Victory* of 1737 and is drawn on vellum. The unknown artist shows the ship as a model with conventional Admiralty Board framing.

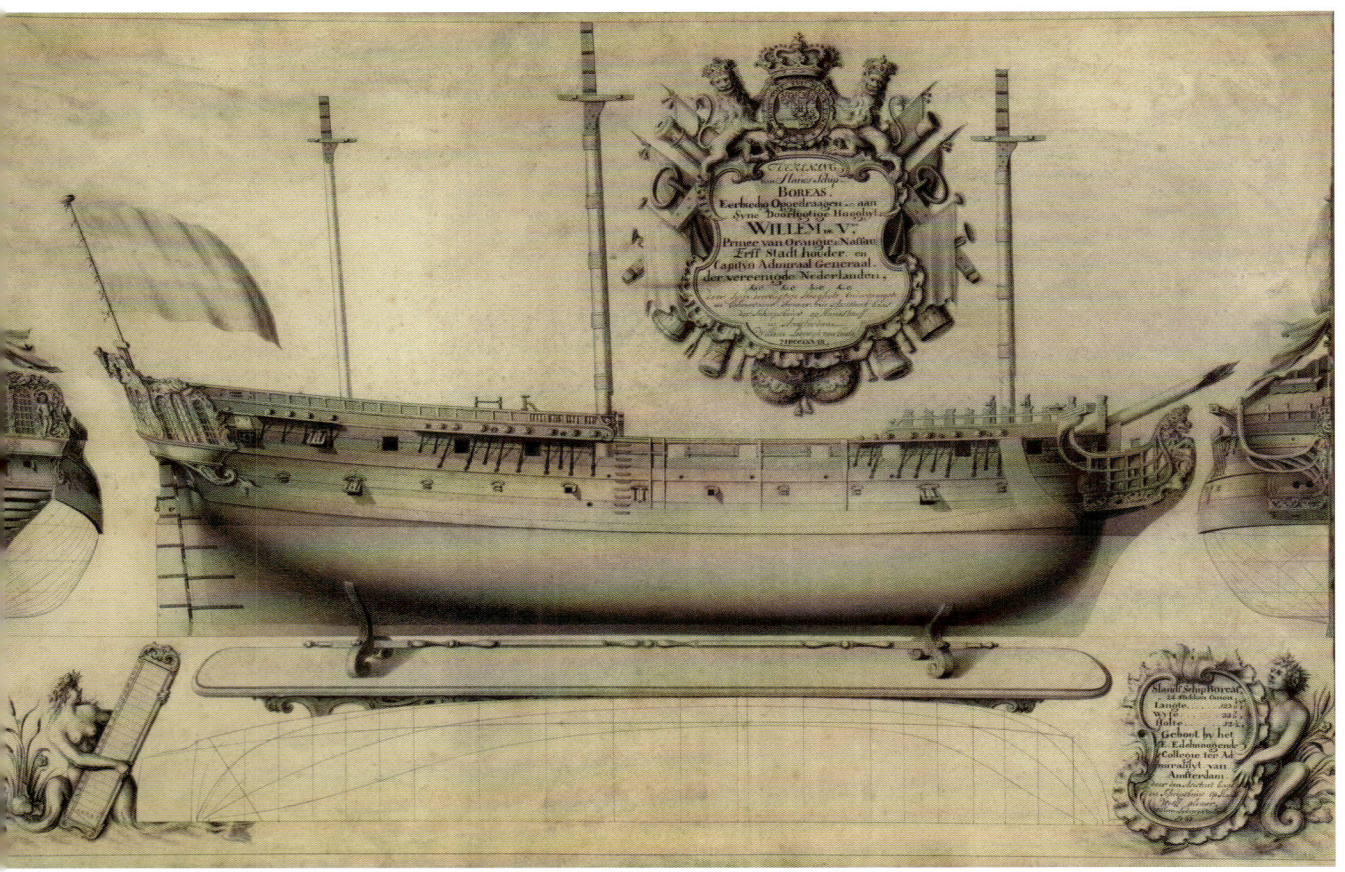

SHIP MODELS IN PERSPECTIVE PAINTED ON PANELS

Two small drawings on paper (7½in diameter) of the same model of an unidentified eighteenth-century three-decker with an unusual female figurehead. The artist is unknown, but the subject was popular enough that he drew at least these two versions.

Below: A broadside depiction on paper of the 1st rate *Royal William* of 1719 (30in by 14in). The ship is shown as a model with conventional framing and launching flags, as were often fitted on early Admiralty Board models.

The Dutch 68-gun ship *De Liefde* launched in 1661. She served as a flagship for both Tromp and De Ruyter and was lost on the second day of the Four Day Fight, 2 June 1666.

CHAPTER 32
Photojournalism in the seventeenth century

WILLEM VAN DE VELDE THE ELDER AND YOUNGER PORTRAY THE WOODEN WARSHIP

BEFORE THE SCIENCE OF photography, visual recording of scenes and events relied upon the art of the draughtsman. With regard to the European navies of the seventeenth century, we are extraordinarily fortunate that two of the most talented artists that ever lived chose to make these wooden warships the focus of their attention. The father and son team of Willem van de Velde the elder and younger have left us an incomparable collection of paintings and drawings depicting these warships in exacting detail. Much of their prodigious output is preserved in museums all over the world, but important examples still remain in private hands, and we have acquired a number of ship portraits by both father and son over the years. Without their efforts, our knowledge of seventeenth-century naval architecture would be greatly impoverished. Their two-dimensional eyewitness images complement the three-dimensional contemporary models to give us a remarkably complete record of the appearance of these awesome war machines.

Willem van de Velde 'the Elder' was born in Leiden in 1611, the son of a seaman. His father was master of a transport vessel and as a young man, Willem followed in his father's footsteps and worked as a seaman until at least 1629. On 19 August 1631, he married Judith van Leeuwen. In October 1632 the couple had their first child, a daughter named Magdalena, and in August the following year their first son, Willem. By 1636 the family had moved to Amsterdam and their second son, Adriaan, was born there that year. The elder's earliest dated drawing was made in 1638, of a ship at anchor, and he continued to draw ships until his death in 1693.

The First Anglo-Dutch War began in May 1652, and the elder van de Velde became involved as both a seaman and a draughtsman. Two days before the Battle of Scheveningen, he was in his own galjoot carrying letters to Admiral Tromp, and he produced eyewitness drawings of both the Dutch preparations and of the battle itself. He continued to document naval actions until the end of the war, and then focused upon producing elaborate grisailles for patrons in Holland and abroad. These grisailles are black-and-white paintings on panel or specially prepared canvas, done primarily with pen and ink. This technique was perfected by van de Velde, and his most ambitious works are among the great glories of Dutch seventeenth-century art.

In 1660, van de Velde was present when Charles II ended his exile in Holland and returned triumphant to England. Van de Velde accompanied the fleet to Dover. With the outbreak of the Second Anglo-Dutch War, van de Velde was in Holland and officially employed to record major naval events for the States General of the United Netherlands. For example, as the Dutch were gathering forces to face the English in what was to become the Four Day Fight in June 1666, the following instructions were issued by Admiral de Ruyter himself,

This is a portrait by the Elder of the English ship *Triumph* built in 1622 and shown as she appeared c1650. The inscription states the *Triumph* 'from before', ie a bow view.

This bow view depicts a large Dutch two-decker from the mid-seventeenth century. It has not yet been positively identified since the stern carvings, which contain the most specific features, are not visible. Nevertheless, comparison with the drawing of the *Triumph*, an English ship viewed from the same perspective, illustrates some of the characteristic Dutch features. These include the unadorned head rails, shape of the beakhead bulkhead and low-slung quarter galleries.

Govert Pietersz, master of the galjoot under his command is hereby ordered to receive on board Willem van de Velde, shipsdraughtman and to take him ahead, astern, or with the fleet or in such manner as he may judge expedient for him to make his drawings … Given on board the States ship the 'Zeven Provincien' under sail in the North Sea.[1]

The Treaty of Breda ended the Second Anglo-Dutch War on 10 August 1667, but the third war followed, officially beginning in March 1672. Van de Velde continued in his familiar capacity as official chronicler of events, and before the battle of Solebay another order was written on 27 May 1672 that reads,

The deputies and plenipotentiaries of their High Mightinesses the States-General of the United Netherlands on board the States fleet hereby direct and order with the advice of Lieutenant-Admiral De Ruyter … that a galjoot of which the master is called Jan Lelij, the galjoot Hollandia, shall take on board the person of Willem van de Velde ships draughtsman and go with him ahead, astern, and with the fleet wherever he may judge it expedient to

This is a portrait of the Dutch ship *Caleb* c1658. Van de Velde has inscribed the name of the vessel on the drawing, but its identity can be independently confirmed by the taffrail effigy and associated stern decorations, which are based upon the Biblical account. It is interesting to note the round object attached to the topsides at the stern above the quarter badge, which is a spare top for the mizzen mast.

make his drawings; and when this has been done, to come and report and await further orders.[2]

The splendid results of being 'embedded' with the Dutch navy are a series of drawings and grisailles for which the art world owes van de Velde a debt of gratitude. These works owe as much to van de Velde's courage as to his talent, as is suggested by an inscription on one of his drawings of the Battle of Solebay preserved at Greenwich that shows a galjoot crossing the bow of a Dutch ship, and van de Velde has written, 'My galjoot trying to bear away to leeward to get quickly out of the way of the engagement.'[3]

In June 1672, Charles II issued a declaration formally inviting Dutch immigrants to move to England. This was a tempting offer as one month earlier French forces under Louis XIV had crossed the Rhine and entered Holland. By autumn, Amsterdam was imperiled and saved only by the opening of dykes around the city that flooded the countryside. The van de Veldes found it difficult to work under these turbulent conditions, and that winter father and son left Holland and settled in England.

The French invasion of Holland was not the only factor motivating the elder van de Velde to relocate to England. In September 1672, van de Velde was defending himself against his wife's accusations that he had been conducting an affair with a married woman, which was evidently true. Earlier, in 1653, the

This drawing by the Elder shows the *Richard*, a Commonwealth ship built by Christopher Pett at Woolwich in 1658. After the restoration her name was changed to the *Royal James*.

elder van de Velde had admitted to fathering two illegitimate children, one by his maidservant and the other by a friend of hers. Evidently, artistic ability and morality are not closely linked on the human genome. His wife did not accompany him to England, but they did reconcile, and in 1674 he returned to Amsterdam expressly to bring her back with him.

The younger Willem van de Velde also experienced marital problems. He wed Petronella le Maire in March 1652, but the following year they separated. Van de Velde the younger is believed to have studied with the great marine artist Simon de Vlieger, and when his marriage was dissolved in 1653, de Vlieger appeared as a witness on his behalf. In 1666, the younger van de Velde married again, this time to Magdalena Walraven.

Upon arriving in England, the van de Veldes did not skip a beat despite their switch of allegiance, and almost immediately found patronage from the English King and his brother James, Duke of York and Lord High Admiral. The elder van de Velde was a witness to both Battles of Schooneveld in the summer of 1673 from a ketch provided for him by the Duke of York. Later that year he accompanied the King and the Duke on visits to Sheerness, but when the final battle of the Third Anglo-Dutch War was fought on 11 August 1673, van de Velde was not there as the King had forbidden it. 'His Majesty was not desirous of him risking his life through cannon fire.'

During the first years that the elder and younger van de Veldes lived in England, they worked exclusively for their royal patrons. The King was so pleased with their work that in February 1674, he instructed the Treasurer of the Navy,

PHOTOJOURNALISM IN THE SEVENTEENTH CENTURY

A drawing by van de Velde the Younger showing the 48-gun fourth rate *Assistance* following her rebuild in 1687. The old *Assistance* was literally falling apart before she was rebuilt by William Castle at Deptford yard in 1687. She was rebuilt three more times and survived until 1745.

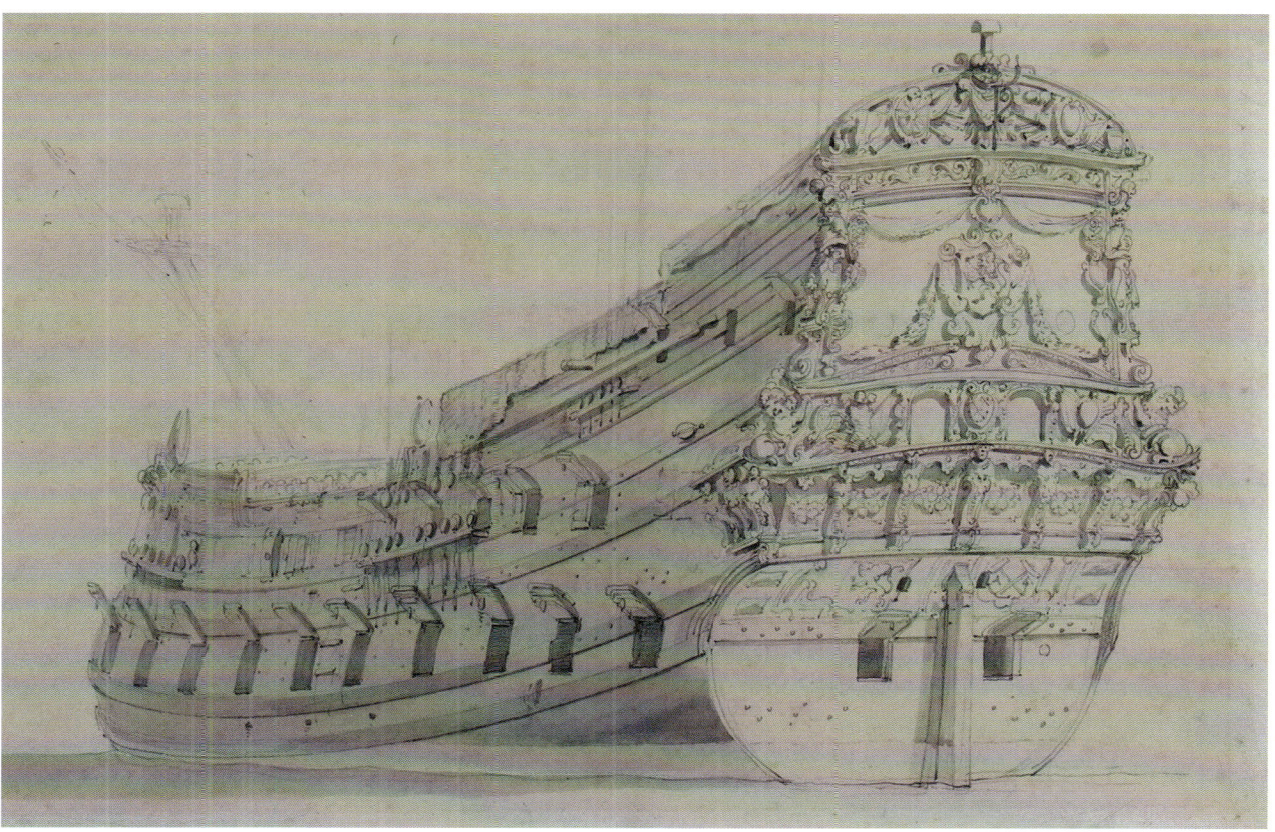

Portrait of the *Vereenigde Provincien*, a 48-gun Amsterdam ship of 1665 that saw action at Lowestoft and the St James's Day Fight.

Whereas Wee have thought fit to allow the salary of One Hundred pounds per annum unto William Vandeveld the elder for taking and making of draughts of Sea Fights, and the like Salary of One Hundred pounds per annum unto William Vandeveld the younger for putting the said draughts into Colours for Our particular use, Our will and Pleasure is, And Wee do herby authorise and require you to issue your Orders for the present and future establishment of the said Salarys to the aforesaid William Vandeveld the elder and William Vanderveld the younger, to be paid unto them and either of them- during Our Pleasure.[4]

The Duke of York added £50 a year to the Elder's income, and this was in addition to what they received for each painting.

The two van de Veldes continued to paint and draw ships and maritime scenes throughout their long lives. Their work was always based on careful observation, and even in 1694, after the Elder's death, the younger van de Velde continued to officially chronicle naval events in the manner established by his father decades before. In familiar language, an Admiralty order dated 18 May of that year reads,

Whereas Mr William Vande Velde is appointed by this Board to goe aboard their Mat. Fleet this summer in ordr to make from time to time Draughts & figures or Imitations of what shall pass & happen at sea by battle or fight of the Fleet, you are therefore hereby required and directed to cause him the said William Vande Velde & one Servant to be born in victuals only on Board Such Ship or Ships of ye said Fleet as he shall desire to proceed in, and that he be accommodated with Such Convenience as can be afforded him for ye better performance of this service.[5]

The younger van de Velde continued to paint until at least 1705 and died on 6 April 1707, at the age of seventy-four.

Both the father and son enjoyed the highest regard an artist could achieve in the age of the old masters, and this esteem has continued without diminution to the present day. Early in their careers, their talents and industry earned them a spot in the highest

Van de Velde the Elder often made rubbings of his drawings of ships, possibly as a resource for use in his paintings. This is the Amsterdam ship *Leeuwarden*, a 36-gun ship of 1645 of which he made several other drawings. This drawing, once the property of the naval historian C R Boxer, appears to be an offset from the drawing of the ship in the Boymans Museum (MB 1866/T248).

echelon of the Dutch marine painters, and later in their careers, after moving to England, they established a studio where they helped train the first generation of British marine artists. The van de Velde studio employed Cornelis van de Velde, who was the son of van de Velde the younger, and J van der Hagen, another Dutch immigrant, along with Isaac Sailmaker and others as assistants. Interestingly, Cornelis later married van der Hagen's daughter, Bernada. One has the impression of a close-knit workplace where Dutch was probably spoken more often than English. The unique vantage enjoyed by the van de Veldes, spanning both sides of the Anglo-Dutch wars and getting up close and personal with so many of the century's most beautiful and powerful warships, enabled them to leave a legacy that is unique in the history of human achievement. All lovers of marine history and art are deeply in their debt.

Interestingly, father and son are both buried in Christopher Wren's lovely St James's Church in Piccadilly. Wren liked this church best of all his London churches, and it features a beautiful font carved by Grinling Gibbons. Situated within walking distance of both Christie's and Sotheby's, this tranquil church has provided a comforting interlude on many of our hectic London visits. The coffee house on the premises provides a lovely antidote to the damp London winters and is highly recommended.

PHOTOJOURNALISM IN THE SEVENTEENTH CENTURY

This is a grisaille or pen painting, showing the Dutch ship *Gouda* at sea with the Dutch fleet before the Four Days' Battle, 1666. It is painted by van de Velde the Elder, and dated 1688. Van de Velde's galliot is seen in the centre foreground. It is interesting to compare this painting to the *St Andrew* picture produced nearly ten years earlier. The compositions are almost identical and clearly related. This observation offers an insight into the art of the van de Veldes, whose compositions generally appear specific and accurate, but may in reality be carefully arranged. The Elder was evidently comfortable repeating a successful composition.

References and Sources

Cordingly, David. *The Art of the Van de Veldes* (London: National Maritime Museum, 1982), pp. 11–20.

Robinson, Michael S, *Van de Velde Drawings in the National Maritime Museum*, Vol. 1 (Cambridge: Cambridge University Press, 1958), p. 451.

Robinson, Michael S, *The Paintings of the Willem Van de Veldes*, Vol. 1 (London: National Maritime Museum, 1990), p. 566.

Right: This grisaille on canvas has suffered over the years, but apparently shows a British three-decker in light airs with a ketch-rigged yacht astern. It bears a signature of van de Velde the Elder and a date that appears to be 1693. Van de Velde the Elder died in December 1693, and this may be one of his last paintings.

HISTORIC SHIP MODELS

Above: The *St Andrew* is shown at sea in the company of other ships. She was a 1st rate, built at Woolwich dockyard and launched in 1670. At the Battle of Solebay in 1672, she served as the flagship of Sir John Kempthorne, and also fought at the battles of Beachy Head and Barfleur in 1690. Painted c1679.

Right above: This is van de Velde the Elder's portrait of Peter Pett's *Sovereign of the Seas* built in 1637. She was the largest, most ornate and expensive ship of the seventeenth century. Built largely to enhance the prestige of King Charles I, her decorations were designed by Thomas Heywood and Sir Anthony van Dyck and covered in gold leaf. An interesting feature of this drawing is the depiction of the gangway leading to the entry port, which reflects her role as a tourist attraction in peacetime. Samuel Pepys records a visit to the ship, during which he placed his wife and four other women in the great Stern lantern and kissed them all and 'were exceeding merry ...'

Right below: The van de Veldes were uniquely positioned to document the British Navy in the latter half of the seventeenth century, but Pierre Puget was an equally skilled French artist who drew detailed views of French ships of the same period. Because Puget was employed in designing and carving the decoration of ships built at the naval dockyards of Toulon and Marseille, his drawings of seventeenth-century French warships are extremely accurate. This drawing on vellum depicts a galley and two men-of-war at a Mediterranean port c1650 (15¾in by 9½in).

PHOTOJOURNALISM IN THE SEVENTEENTH CENTURY

This is a portrait of the *Royal Charles*, a 1st rate built at Portsmouth by Anthony Deane in 1673. She was Prince Rupert's flagship at the first battle of Schooneveld on 28 May 1673. This painting, dated 1676, was painted as a companion to the portrait of the *Royal James,* and it is likely that both were commissioned by Anthony Deane himself.

Portrait of the *Royal James,* a 1st rate built at Portsmouth by Anthony Deane in 1675 and painted in 1676. This painting and its companion the *Royal Charles* were likely commissioned from van de Velde by Sir Anthony Deane as they represent the two 1st rates built by him.

This portrait of the *Britannia* built by Phineas Pett at Chatham in 1682, shows both a stern and broadside view of this handsome 1st rate. This painting is by Isaac Sailmaker and was most likely painted while he was working in the van de Velde studio as it is based upon a broadside drawing of the ship by the Elder (now in the collections of the Royal Museums Greenwich).

Two small Napoleonic prisoner of war bone ship models c1800 on their original plinths (12in and 11in long). The hulls are planked in bone strips with horn or baleen wales. The rigging is made of spun silk and is largely original on both models.

CHAPTER 33

Napoleonic-era ship models

Our interest in model ships derives mostly from an aesthetic perspective. It is the beauty in the shape and decoration of these wooden warships that appeals to us most. Unfortunately, these aspects of ship design eventually gave way to practical considerations, and by the nineteenth century the tumblehome, sheer and decorations that characterised the earlier ships began to disappear. For this reason, our interest in collecting builders' models ends with the conclusion of the Napoleonic Wars.

Navy Board ship models embody the entire process of design, construction and embellishment of warships of the golden age of sail. They are superb achievements of craftsmanship and are imbued with the character and conventions of their time. However, they only indirectly convey the romance, struggle and depredations of the sailors and their commanders who animated them when at sea. The opposite can be said of Napoleonic era prisoner of war models. These models, often highly detailed beautiful achievements, were produced by prisoners living in dreary if not deplorable circumstances working with primitive hand tools and using near worthless materials. Many hundreds still survive.

Our interest in PoW ship models spans many decades, and we have acquired most of ours at public auctions in the US, UK and France. The dealers' trade in these objects, however, was dominated in the late twentieth century by two establishments. One was the Parker Gallery, the oldest established firm of picture and print dealers in London, which was run by Bertram Newbury, succeeded by his son, Brian; and the other was Langford's Marine Antiques, which was founded and run by Laurence Langford, assisted by his wife, Jane. Langford's originally operated out of a small shop on Chancery lane, near London's silver vaults. Alas, neither of these venerable shops exist today. Over the years Laurence Langford became a good friend whose sense of humour never failed to entertain, even when he outcompeted us for a model. I (Arnold) recall an incident that occurred when I was on sabbatical at Oxford, long before the internet brought every obscure auction house sale to our attention. I came across a notice in a regional newspaper that announced a routine sale of household effects taking place in Glasgow the following week that included a small boxwood PoW model misidentified and grossly undervalued. I made arrangements to fly to Scotland for the sale, confident that I would return with the model in hand. I arrived in time to view the model one hour before the auction. As I was intently examining the beautiful little model, I heard a familiar voice behind me: 'A lovely model ain't it Arnold?' It was Laurence Langford. I returned to Oxford with some marine books but the model wound up in a small shop on Chancery Lane.

ORIGINS

The unlikely origins of these remarkable works of art have captured the imaginations of collectors for over 200 years. Detailed

The finely carved stern of this small bone PoW model is relatively accurate for a 74-gun ship of the Napoleonic period. Note the cords with pulls leading into the stern chase ports. When they are pulled the broadside guns retract.

A small bone PoW model of a 74-gun ship. The unusual figurehead appears to be a winged warrior. A pair of ship's boats are suspended from the yardarms.

descriptions of the industry of ship model making among French prisoners of war housed in England during the Napoleonic wars can be found in the works of Ewart Freeston[1] and Clive Lloyd,[2] and we strongly recommend these to interested readers.

Throughout the Napoleonic wars there was a shortage of sailors to man the French fleet. One solution was to press able-bodied conscripts into the Navy, and prime candidates were men living in coastal towns. The town of Dieppe located on the

Most PoW models share similar features including the deck arrangement. As shown in this example, the foredeck is fitted with a jeer capstan, a stove pipe and there is a belfry incorporated into the forecastle bulkhead. Stylised carved hammock nettings line the upper deck along the gangways. A main ladderway leads to the poop deck; this is sometimes fitted with a landing, though not in this example.

Normandy coast across the English Channel from England was well situated to help meet the demand. Dieppe, a prominent trading port, happens to also have a tradition of ivory carving that began in the renaissance when trade between Dieppe and Africa provided a ready source of elephant tusks. Workshops in Dieppe became famous for finely made, quite expensive, carved ivory objects, including ship models. When Napoleon Bonaparte visited Dieppe in 1802, he acquired a model of a 74-gun ship, which he presented to his wife, Josephine.[3] These objects were marketed to the aristocracy. However, demand was dramatically reduced after the French Revolution, and when the Napoleonic wars began scores of ivory carvers were forced into the Navy. One can only assume that these craftsmen were poor sailors, and most likely were numbered among the crews captured in huge numbers by the Royal Navy during the early years of the war. We think it is likely that the French ivory carvers from Dieppe, initially imprisoned in hulks and later in depots on land, initiated the practice of carving trinkets out of bone to sell to the British public. They may well have trained fellow prisoners in the craft. The sheer numbers of models indicate that many artisans were at work, and based on the variable quality of

NAPOLEONIC-ERA SHIP MODELS

An unusually fine bone model bearing the name of the famous frigate *Pallas* at the stern. The hull is neatly planked and pinned with bone and horn strakes. This model once belonged to the author Kenneth Roberts and features in his novel *The Lively Lady*, whose protagonist Captain Richard Nason of the American navy is captured by the British in the war of 1812 and sent to Dartmoor Prison.

elements in some models, teams of craftsmen may also have been employed.

So many sailors were captured by the Royal Navy that there was a shortage of space to house them. Decommissioned and captured warships were dismasted and converted into prison hulks in harbours along the coast, including at Chatham, Plymouth and Portsmouth. The largest hulks could hold up to 1,200 men. From personal accounts by prisoners we learn that items including gambling pieces, portraits, landscape paintings, straw work items and bone ship models were made by prisoners in the hulks.[1] However, there is abundant evidence that workshops producing ship models were also found in the prisons and depots on land. Throughout the Napoleonic Wars prisoners were housed in prisons and castles across Britain. The first and most famous depot purpose-built to house Napoleonic war prisoners was built at Norman Cross near Peterborough in 1797. It housed 7,000 prisoners when it opened in 1798, though it is worth noting that there were a total of 35,000 French captives in Britain that year. Four to five prisoners were assigned to cells measuring 8ft by 7ft,

Three small bone PoW models. The example on the left has a solid hull fashioned from a bovine tibia, the other two are planked in bone strips over a hollowed wood core. They are all mounted on original baseboards and the centre example is preserved in its mirrored straw work case.

A miniature bone model of a three-decker just 4½in long. Despite its small size, the model has a planked hull with fittings usually found on larger models including: rudder chains; wales made of horn; and gudgeons, pintles and mast hoops made of gold foil. The baseboard indicates that the model was originally enclosed in an octagonal case.

and larger rooms of 24ft by 22ft accommodated up to forty-five prisoners divided among nine cells. In 1806 another large prison was built at Dartmoor to relieve the congested Milbay prison at Plymouth. It housed 6,000 prisoners when it opened in 1809. A significant number of fine ship models were produced at these two depots as attested by labels surviving on some of them. Models were usually sold at markets held at the prisons and the model makers could earn sizeable sums. A market was held at the Norman Cross Depot from 10am to noon, either daily or twice per week, where prisoners set up stalls to sell their handicrafts. Reverend A H Davis, who visited the market in its heyday, remarked that 'more than 200 pounds a day has been frequently paid out in purchasing

PoW models were often housed in straw work cases, of which this is a typical example. They usually had a mirrored interior to allow viewing of all sides of the model. This 13in-long bone model of a three-decker has a planked hull and horn wales and retains much of the original rigging.

their labours'.[2] A small bone ship model 12in long was sold for £26 by a prisoner at Portchester Castle, and at Forton prison, near Portsmouth, a bone model that took two prisoners over six months to make was sold for over £40.[3] Remarkably a sailor named Garnier from St Malo spent one year making a model 2in long that he sold for the princely sum of £100. In the Dover Museum there is a 2ft-long bone model of the *Cesar*, a 74-gun ship, alleged to have been made in Dover Castle by a French prisoner. A story associated with this model claims that the dedicated modeller stayed in England following the Peace of Amiens in order to finish the model, leading to his reincarceration when the war resumed one year later.[4]

HISTORIC SHIP MODELS

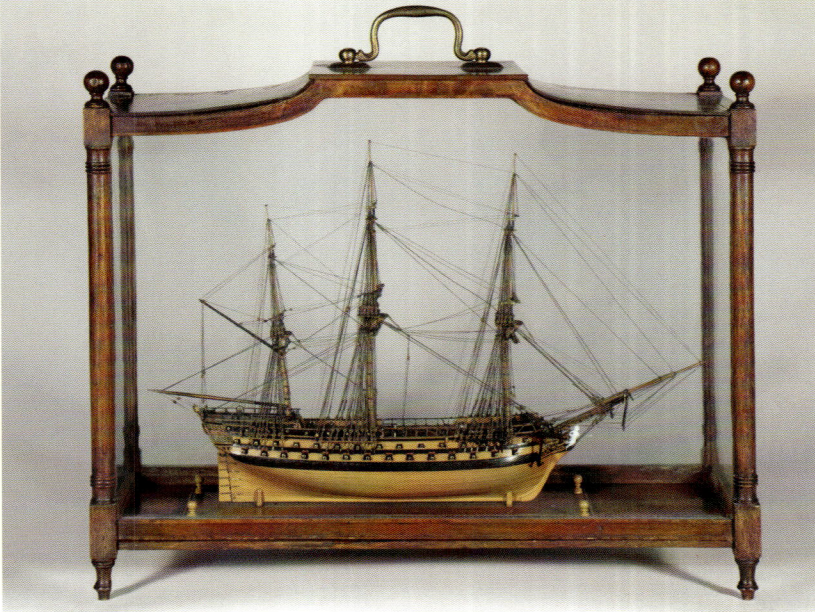

Two very fine boxwood models depicting a *Diana* class frigate (13¼in long) and a 74-gun ship. These models of the Napoleonic period were made by the 'Pilkington maker', the same craftsman who made a set of similar boxwood models that were part of a gift from Dame Mary Elizabeth Pilkington to the Liverpool Maritime Museum in 1921. These models have much in common including their accurate lines and proportions, the extensive use of boxwood and ebony, and the use of ivory for the figureheads, stern carvings and deck details. It is likely the maker was working from accurate drafts. The hulls are finely planked with scale-size strakes. The decks are similarly planked with tiny trenails, and in a remarkable feat of craftsmanship, cabins with bulkheads and glazed windows are fitted throughout, including stern cabins beneath the poop decks that are essentially invisible. Interestingly, most of the models are housed in similar attractive display cases as seen in these two examples.

CONSTRUCTION

Materials

Materials used in the construction of ship models and their cases included wood, bone, horn and straw. Bone models are almost always made from cow bone, presumably acquired as left-over scraps from the kitchens. Wooden models are made from a variety of woods, with the finest miniature examples often constructed from boxwood, and some have a copper-sheathed hull. Auction houses and galleries often refer to the darker wales and inlays used in bone models as being components made of baleen. While this is the case for some, in our experience many are cow horn selected for its dark colour, either brown or near black. Metal fittings including guns, anchors and ringbolts, were usually made of brass or steel, and mast hoops and rudder pintles and gudgeons are surprisingly often simulated in gold foil. The rigging is usually silk, spun and often cable laid as in full-size practice, though some models are said to have been rigged in human or horse hair. Baseboards and cases are often decorated with straw inlay.

CHARACTERISTICS

Unlike Navy Board or builder's models, these are obviously not scale models. The accuracy of the hull shape and proportions, the length of the bowsprit, the height of masts, the relative size of fittings, etc. are most often incorrect. Wooden or boxwood models usually have better proportions than the bone models and are often more detailed. The rigging on both bone and wooden models, if original, is almost always highly detailed and, in most cases, accurately reflects contemporary, albeit French, practice. This is probably because the builders were most often sailors themselves and so had first-hand experience with the rigging of ships. On the other hand, the models were almost always over-rigged and over-gunned. They most often have every possible spar displayed, including topgallant masts and royal yards, overlong bowsprits, and feature the liberal use of studding-sail booms, even on the mizzenmast, topgallant and royal yards where they were almost never used.

The stern of the 'Pilkington maker' frigate incorporating miniature ivory carvings that are quite impressive since the entire stern measures just 1½in across.

A small boxwood copper hull model with fine carved decorations on the taffrail, forecastle breastwork and along the frieze planking.

The bow of the 'Pilkington maker' two-decker with a lovely ivory figurehead of a warrior holding an extremely delicate spear.

The general arrangement of ships' fittings is invariably French. A fore jeer capstan on the fore deck, a main ladderway with a berthing or landing descending from the poop deck, and a 'horseshoe'-shaped stern are French features that are almost universal on PoW models. Nonetheless, a model of a French ship would have had little appeal to British citizens at the time, so the flags or pennants are British, as are the ships' names that occasionally appear at the stern.

The French prisoners were also fond of carving animated toys and amusing automatons out of bone. These included cleverly operated figures of Breton women seated at spinning wheels, often with accompanying figures who dance or rock baby cradles when a wheeled mechanism on a tier below is turned by a hand crank. Toy guillotines were popular and had a mechanism whereby the executioner released the dreaded blade to decapitate a victim, invariably female, and presumably an effigy of Marie Antoinette. Similarly, ship models were often animated. A spring hidden inside the hull would enable the guns to retract when a string was pulled. Two separate cords allowed either port or starboard broadside to be activated separately.

TOOLS

There is abundant evidence that very fine metal hand tools such as scrapers, files and chisels were the primary tools used by the model makers. Marks on carved bone elements often bear scratches left by files. Turned elements such as guns, masts and some rigging blocks indicate the use of lathes, assumed to be treadle or relatively simple miniature spring pole lathes. Ewert Freeston describes a jeweller's Archimedean drill that can easily drill very fine holes using needles ground to a triangular cross-section.[1] It is difficult to imagine how the extremely fine detail on the smallest miniature models was achieved without the aid of some form of optical magnification.

Historical Perspective

THE SHIP MODEL TRADE FLOURISHES DURING THE NAPOLEONIC WARS

The passion for collecting miniature ship models during the Napoleonic era in Britain was not confined only to those works

A finely detailed small boxwood PoW model with coppered hull and colourful deck details. Note that there is a watch bell at the quarterdeck breastwork in French fashion.

Two examples of extreme miniature models made entirely of boxwood including boxwood sails and rigging. The example at left is 4½in long, the one on the right is only 3in in length. Miniature boxwood models such as these usually included

elaborately fashioned bases as seen on these examples. An old inscription on the bottom of the case of the smaller model states that 'this ship was much admired by Prince Albert when in Bristol'.

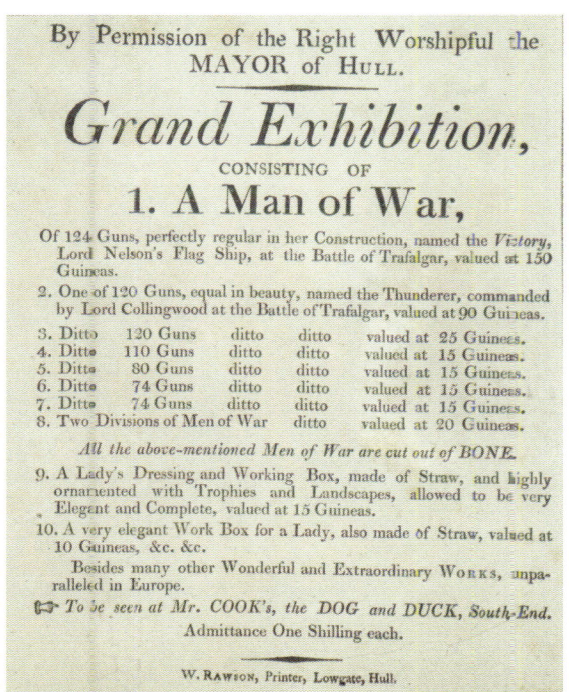

A rare Napoleonic-era broadside advertisement for a sale of PoW ship models 'cut out of bone'. It gives an idea of the value placed on these models before they became antiques.

A contemporary painting of the prison hulks in Portsmouth harbour, by Ambroise Louis Garneray. Garneray was a twenty-three-year-old sailor on the frigate *Belle Poule* when it was captured by the British in 1806. He was imprisoned in the prison hulk *Prothee*, and before his release in 1814 he painted a number of views of these floating prisons.

produced by prisoners. Wealthy patrons sought out microminiature ivory carvings of warships made by a pair of artists working in Bath and London between 1790 and 1804. G Stephani and J Dresch were originally from Augsburg and they operated a successful workshop whose customers included King George III. They were frequent exhibitors at the Royal Academy of Arts in London, and their most popular composition featured a frigate with other vessels sailing on an ivory sea, sealed in a miniature convex Bristol glass case.

There is another noteworthy set of models produced during the Napoleonic period in Britain that have been attributed to prisoners, but we feel wrongly so. These are accurate and beautiful models made of boxwood with ivory trim and often contained in distinctive wooden and glass carrying cases. A set of over a dozen were found in a house in Dublin in the early twentieth century and purchased by Sir George Pilkington, whose family donated them to the Merseyside Maritime Museum. We believe that they were found in the home of the unknown but gifted man who had made them over 200 years ago. He made models of all sizes of sailing craft, and they all appear to be accurate scale models. Several models that are clearly by the same maker have turned up at auction over the last few decades, and we were able to obtain two of these, illustrated in this chapter.

Legacy

Ship models made during the Napoleonic Wars have been collected and admired for over 200 years. Writing in the 1920s, the renowned model builder and restorer Henry Culver described their appeal in the following way:

> Things of beauty, the carven ships are a joy that lasts forever. They bring to us a thrill from out of the past; they tell us of romance, of struggle, of suffering, of patriotism. They are the guerdon of the dying art of a past epoch. And if they bring to some of us today these sensations, rare in our commonplace lives, who shall say that good has not come out of evil?[4]

This frigate is an ivory microcarving made by G Stephani and J Dresch in their London workshop c1800. It is set into a concave piece of Bristol glass simulating a blue sky and is covered by a clear convex glass lens, which creates a little magnification and gives the diorama an ethereal, jewel-like quality. It is only 3¼in high and comes in a tooled leather carrying case.

The model of the *Diamond* had active woodworm infestation and was gently treated by anaerobic immersion in argon gas for four weeks to destroy the insect larvae. Several old worm holes are visible surrounding the carved merman figure. These are exit holes and mark where wood-boring beetle larvae abandoned ship after consuming their fill of boxwood.

CHAPTER 34
Care and conservation

The models in our collection have survived for over two centuries, and in some cases, well over three centuries so far, and with care they should survive well into the foreseeable future. Because the models are made of organic materials and metals including wood, brass, copper, linen, silk, shellacs, varnishes and organic pigments, they are subject to degradation and change over time, and therefore do require some curation.

AVERTING THE RAVAGES OF DUST, LIGHT, HEAT, DAMP, DRYNESS, INSECTS, TRAUMA AND CONSUMPTION

Maintaining a collection like ours requires attention to basic conditions of heat, humidity and light, as would any collection of fine art or antiques. Naturally, the models are all kept in display cases, and we only use glass, not plexiglass or Perspex, in order to protect them from curious fingers as well as dust. They are placed out of direct sunlight and away from heater gratings or the direct path of hot or cold air vents in order to minimise abrupt temperature changes. All windows are treated with UV screening film, but non-UV light can also fade objects over time. The degree of light screening has necessarily been a compromise between the gloomy dark that we prefer for the models and the bright sunny rooms favoured by our wives. Humidity, or rather, dryness, is a major threat. Too much damp will encourage the growth of mould, but dryness can shrink, split and warp wood. Ideally, the models should be kept in environments of constant humidity of around 50 per cent. This is also probably true for most people. In practice, this is almost impossible to achieve. We have been content if the winter humidity ranges no lower than 35 per cent and summer no higher than 55 per cent. When both of us lived in New England, this meant installing humidifiers in the central heating system and hygrometers in the living rooms to monitor what was often a temperamental system. Central air conditioning in the summer months adjusted to keep us comfortable and served the models equally well. We have monitored the condition of the models over decades now and everything appears quite stable.

From time to time we have had to deal with active woodworm eating away at centuries-old fruitwood timbers. Woodworms are not worms but the larvae of the common furniture beetle *(Anobium punctatum)*. The adult beetle lays eggs on the wood surface, and the newborn larvae burrow in and tunnel along inside, generally along the grain of the wood, for five to ten years. The larvae then approach the surface and create small pupation chambers, where they spend six to eight weeks before becoming adults and emerging. The emergence holes are 1–3mm and are the most obvious sign that woodworms have been present, but not necessarily that they are still active. Actually, the presence of at least some woodworm holes in a model is a reassuring sign that the model is a genuine antique, and by the time the exit holes appear, the beetle has already emerged. In fact, if one suspects fake worm holes, as can be produced by shotgun pellets, for example, inspection of a genuine worm hole under magnification should reveal the slightly larger pupation chamber just below the surface. The trouble is that, once emerged, the beetle may lay eggs within as little as two weeks, and the process will restart. The fresh appearance of tiny piles of sawdust (actually the excreta of digested wood known as frass) indicates an active infestation, and we have encountered this problem with one of our models. There are a variety of treatments available to deal with woodworms, but most involve sprays, soaking, paste, etc, and are not suitable for models. Many years ago, it was possible to fumigate a model by placing it in a vacuum chamber and infusing a highly toxic gas, methyl bromide or ethylene oxide, to kill the larvae. Now however, the use of methyl bromide is no longer allowed in most countries. We settled on a more environmentally friendly approach. We hired a firm that treated the problem *in situ*. The model was enclosed in an airtight, low-gas permeability plastic bag, and the air inside the bag was purged and replaced with argon, an inert gas. The humidity was adjusted to 50 per cent, and the bag was heat sealed and kept sealed with an oxygen tension below 0.1 per cent for four weeks. This treatment method is reported to eliminate mould and fungus as well as insect infestation and was entirely safe for the model. It also had the added attraction that the model was treated at home.

CHAPTER 35
Fakes and forgeries

IN THIRTY-FIVE YEARS of collecting, we have encountered three 'modern' models that were alleged to be period Admiralty models.

THE GOOD, THE BAD, AND THE PHONY

One was a model of a seventeenth-century three-decker but turned out to be based upon the 2nd-rate *St Michael* model at the National Maritime Museum in Greenwich. Perhaps in an attempt to make it appear more like an original design, its builder incorporated a few deviations from the prototype. Principally, this included substitution of a lion for the equestrian figurehead. Henry saw the model around 1985, making a special trip to the Gulf coast to examine it in a private collection. While the overall look is fairly convincing from a distance, and the size and materials are correct, on close inspection the model is a clear reproduction. It has no signs of any losses or repairs, is complete, and has no signs of age. In particular, there is no internal dust, no worm or other environmental damage, and most damning of all, the quality of construction and decoration is not up to period standards. The owner had purchased it as an antique admiralty model from an East Coast dealer some years before but had no information about the prior history of the model. If it turns up for sale in the future, it would not convince a discriminating collector.

The second fake model we saw was also in a collection in the southern United States. Also purporting to be of the seventeenth century, this represented a 3rd-rate ship from about 1680. Henry had also made a trip specifically to see this model about twenty-five years ago, and his disappointment began when he first caught sight of the model from across the room. It had a fully planked hull, which is a practice not introduced in British dockyard model construction until well into the eighteenth century. As Henry approached, he became certain that this was not a genuine artefact. Once again, the model was complete and uniform in quality and condition, with no repairs or significant damage. Not only was the quality of workmanship and decoration below period standards, but there were errors in shape and proportion involving major elements of the head, stern and quarter galleries. Materials were incorrect also, as, for example, the use of oak for planking. Carved decorations included anachronisms, with such mistakes as Roman plumed helmets on the knighthead warriors. This model did have some signs of age, including dirt, pigment losses and some warpage, but sitting on a table for several decades would suffice to account for this minimal deterioration. As in the previous instance, the owner had purchased this model from a dealer who claimed it was an authentic seventeenth-century relic, and he was disappointed that we were not interested in purchasing it. Years later, we learned a bit more about the early history of this model when we came across it in an old auction catalogue. It turns out the model had been sold by the American Art Association at their gallery on Madison Avenue in Manhattan on 6 January 1926. It was lot number 847 and was not described as a period Admiralty model at that time.

The third was a model of a 64-gun 4th rate c1660–70. This was a handsome model sold at Christie's in New York in January 2008. We inspected the model carefully and felt it was deliberately made to appear older than it was. It had previously (originally?) been sold by the London firm Arthur Davidson Ltd in 1965. Unlike Napoleonic period ship models that exhibit enough variation in style and quality to make forgeries look fairly convincing, British dockyard models were uniformly built to a high standard and are very difficult to reproduce, even by skilled craftsmen using modern tools. They require great artistic talent as well as woodworking skills and historical information that modern model builders rarely possess or have access to. In contrast to other antique ship models, they have always been sufficiently rare and valuable so that some provenance or history of the model is usually available. For these objects, a convincing forgery is nearly impossible to make.

CAVEAT EMPTOR

The situation with regard to Napoleonic PoW models made of bone is altogether different. Among these models, fakes abound.

FAKES AND FORGERIES

Many have been made as deliberate forgeries, while others have been made in the style of prisoners' work and subsequently passed off as the real thing. We know of at least two prominent collections that consisted predominantly of twentieth-century models that were sold or donated as authentic antiques. If the reader should have fallen victim to one of these scams, there is, however, reason to take heart. Much to our surprise, these modern creations in the early style often bring prices at auction that are comparable to what the originals would bring. The appeal of these wonderful bone models is evidently great enough that the market does not discriminate against those produced after the Industrial Revolution.

There are no fake models in our collection. However, when we acquired this model it had evidence of considerable 'restoration'. It had been altered many times over the years, and had new decks, bulwarks and deck fittings with many anachronistic and contradictory features. There nonetheless appeared to be an old hull lurking underneath. We carefully removed the obvious newer additions to reveal an authentic early eighteenth-century dockyard model in frame. Authorities at the Royal Museums Greenwich suggest that the model may represent a merchant ship dating from the period of Queen Anne.

277

All photographs show the model as it was when we bought it in October 2020. They are provided courtesy of Aguttes, Maison de Ventes aux Enchères, Paris.

This broadside view is unmistakably that of an English 4th rate of the late seventeenth century. It is pierced for 52 broadside guns and is decorated in the baroque manner with gilded carved decorations. Below the lower wales the frames are clad in thin copper sheeting, and although this may account for some distortion of the profile, the long rake of the stem and sheer of the quarterdeck are obvious. The anachronistic rigging appears to conform to early nineteenth-century practice, save for the presence of a lateen yard on the mizzen mast.

APPENDIX
A William III 4th rate c1695

Acquisition

IN 1926 WHEN HENRY Culver wrote his catalogue of all the known seventeenth-century ship models in the world, there were thirty British Navy Board examples.[1] During the ensuing ninety-four years, twenty-two additional models have come to light built prior to 1702, bringing the total to fifty-two. Of these, twenty-two are in museums and eleven are in private hands, including two yachts and nine men-of-war. Five of the privately owned models are described in this volume. We were therefore astonished to see a previously unknown example appear at an auction in Paris in late 2020.

At the time we were hoping that no one else would recognise this 'new' seventeenth-century model, because it was in disguise. The auction house described it as a 'superb and rare English frigate from the mid-eighteenth century, edged with mahogany …' The hull was clad in copper beneath the lower wale, the decks were fully planked over, and it carried Napoleonic-era rigging. With no open ribbing and exposed deck timbers, it did not really look like a seventeenth-century Navy Board model and could have been mistaken for a model made around 1800. The hull shape and carved gilded decoration, however, was strictly seventeenth century and it even sported a William III cypher on the stern railing.

Coronavirus was raging at the time, precluding a trip to Paris, so we had to rely on photographs to understand what the model truly represented. Between the lower wales one could just discern individual frames butted against one another, convincing us that there was an open-framed Navy Board hull beneath the coppering. We concluded that this was a late seventeenth-century British Navy Board model that had been brought up to date a hundred years later with the addition of 'missing' planking on the decks, copper sheathing of the hull, and new masts, spars, and rigging. These additions were now 200 years old, and we hoped the confusing appearance would put off other potential buyers. The estimate was quite reasonable, but we were prepared in case the auctioneer had been too conservative.

As it happened, we were not the only ones who saw through the 'disguise', and it sold for more than ten times the estimate. We did prevail, however. All of this transpired too late for the model to be included in the body of this book, but with the kind permission of the publisher we are including it in this appendix. It is our sixth seventeenth-century model. As we write this appendix, the model remains in France awaiting an export licence.

The stern is richly carved and gilded. Virtually all of the original carved work appears to have survived. The taffrail is centered by the arms of King William III flanked by the English lion and unicorn. The frieze carving beneath the taffrail is gold leaf on a red ground and features Tudor roses. This theme is carried through to the panels beneath the breast rail of the open gallery, which flank a delicately carved central panel bearing the monogram of King William III. There are two stern ports in the lower counter, as was customary at this time. The quarter pieces are kilted warriors and the panels beneath them feature Tudor roses.

This view of the waist taken from above shows how the deck beams have been covered over by planking scored to represent seams and butts, presumably added when the model was 'updated' c1800. The planking along the sides is original, and the added planking can be distinguished by its discordant colour and condition which is warped and split. The gratings and coamings, however, may be original. The forecastle bulkhead can be seen at left, and the side companionways are an unusual feature seldom seen on Navy Board models.

Description

CONDITION

The original model survives largely intact, with minor losses. These consist of the flanking figures on either side of the lion figurehead, and the midline structures on the decks, including the belfry and main jeer capstan. Additions were fitted around 1800, but these should be easy to remove once we have access to the model. They consisted of thin copper sheeting applied to the ribs beneath the waterline and on the stem, keel and rudder; planking applied over the (originally exposed) deck beams; and anachronistic masts, spars and rigging. A thin layer of varnish was also applied to the hull.

The rigging on the model when we acquired it was consistent with Napoleonic-era practice featuring a jib-boom with dolphin striker, tops with square aft profiles, etc. but there was one glaring anomaly – the presence of a lateen yard on the mizzen mast. This spar would have been quite anachronistic c1800 when the rest of the rigging was applied, but would have been appropriate when

A single stern lantern survives, and it is of typical late seventeenth-century form with mica windows. The lantern brackets are missing. The recessed open gallery at the level of the upper deck can be seen, with its moulded screen bulkhead and red inboard planking. The copper sheathing applied to the hull is also apparent, held in place by small nails. It covers the frames and is presumably part of the Napoleonic period transformation of the model.

the original model was built c1690. It raises the possibility that some of the other yard arms may be original, and therefore, that the model might have been originally rigged. We can think of no other plausible explanation for the presence of this spar. We do not know whether the wood the aged spars are made from or the paint that covers them is 300 years old or 200. The question of whether the model was originally rigged has not been answered conclusively.

A WILLIAM III 4TH RATE c1695

CONSTRUCTION

Scale: 1/48 Hull length: 37¾in

The model has fifty-four gun ports, with none on the forecastle. Several lines of evidence date the model to 1695. The presence of the cypher of King William III on the stern railing means the model could not have been made prior to the death of Queen Mary in 1694. Other carvings on the model are consistent with this date; for example, the vertical terms on the beakhead bulkhead, which characteristically do not reach the deck, a feature that we have only observed on ships from the end of the seventeenth century. The relatively long rake of the stem and round tuck stern suggest that it dates from early in William's reign.

There are some unusual and notable features including the twist-stairs leading down from the quarterdeck and the ladders descending from the fore deck, a feature rarely, if ever, shown on seventeenth-century models. The dolphin-shaped hancing pieces are also noteworthy, as is the exuberant carving over the quarterdeck bulkhead, which displays cherub heads supporting the royal coat of arms. The stern has an open gallery and a taffrail featuring a crowned royal coat of arms flanked by a lion and unicorn. The quarter galleries include crossed batons topped by fleur-de-lis, symbols associated with King William, and two male heads in profile on either side of the single light. The lower finishing is in the form of a winged cherub head. The canopy is carved with floral motifs in low relief. The gilded surfaces appear to be original and have now acquired a lovely soft patina of age.

The quarter gallery survives intact and features a central mica light flanked by male busts carved in relief. Other carvings include trophies of arms and Tudor roses. The lower finishing of the quarter gallery is a winged head of a cherub, which is typical of models prior to the eighteenth century.

The starboard quarter gallery is better preserved than the port side, insofar as it retains a small vase of flowers at the front of the roof. Carving can be seen to extend across the top edge of the taffrail.

The broad area of plank above the main wales is scribed to represent the seams and butts between individual strakes, in accordance with seventeenth-century practice. The channels are braced from above by arched metal spurs, and the short chainplates are single bolted to the lower strakes of the chainwales. The edges of the channels are finished with a metal band that is taken round the ends of the channels. Note that the hances are in the form of dolphins rather than the more common crouching dogs.

Above: The lion figurehead is a handsome example and is one of the earliest that sports a tail. The carved and pierced foliate trailboard echoes the decorative scheme that appears on the stern. The carved brackets on the head timbers are typical of those of the seventeenth century, as are the figures on the cat supporters. Note that the carved brackets on the beak bulkhead do not reach the beak platform – a feature that we believe can be reliably dated to the last decade of the seventeenth century. Red paint, presumably added c1800, is peeling away from the stem revealing the boxwood underneath.

Left: After part of the waist, showing the quarterdeck bulkhead, gangway and stairs. The bulkhead features a finely carved coat of arms of William III surmounted by a crown and surrounded by baroque drapery and cherubs. The circular apertures beneath the breast rail are loop holes for small pieces of ordnance to repel boarders. The doors hung on hinges in the bulkhead have handles, an unusual feature. The winding stairs and long gangway to the quarterdeck are adorned with rich carvings below the handrail.

Chapter Notes

Introduction

1. R C Anderson, *The Rigging of Ships in the Days of the Spritsail Topmast, 1600–1720* (Salem: The Marine Research Society, 1927).
2. Samuel Pepys, *The Diary of Samuel Pepys*, Latham, R and Matthews W, eds, Vol. 4 (Berkeley and Los Angeles: University of California Press, 1970–1983), p. 437.
3. R Morton Nance, *Sailing-Ship Models* (London: Halton & Truscott Smith Ltd, 1924), p. 11.
4. *Some of the Ship Models in the Collection of the Nautical School T.S. 'Mercury', Hamble, Hants* (London: The Avenue Press).
5. 'Keep the "Mercury" Collection of Ship-Models in England!', *Illustrated London News*, 174, 4693 (30 March 1929).

Chapter 1 The *Royal James*, 1st rate of 1671

1. Samuel Pepys, *Samuel Pepys's Naval Minutes*, J R Tanner, ed, Vol. 9 (London: Navy Records Society, 1926), p. 374.
2. Pepys, p. 186.
3. E H Pearce, *Annals of Christ's Hospital* (London: Hugh Rees Ltd, 1908), p. 123.
4. Edmund Blunden, *Christ's Hospital: A Retrospect* (London: Christophers, 1923), p. 62.
5. Wilson records that, 'In a room between the upper end of the Hall and the Mathematical school are correct models of the various sized vessels now composing the wooden walls of old England, with the name of the donors.'
6. Gregory Robinson and R C Anderson, 'The Identification of Models of Men-of-War', *Mariner's Mirror*, 2 (1912), pp. 339–41.
7. J R Tanner, *The Naval Manuscripts in the Pepysian Library*, Vol. 1 (London: Navy Records Society, 1903), pp. 230–1.
8. Howard Nixon, *Catalogue of the Pepys Library at Magdalene College, Cambridge VI: Bindings* London: D S Brewer, 1970, 1984, item 47a and Introduction.
9. Anthony Deane, *Deane's Doctrine of Naval Architecture, 1670*, Brian Lavery, ed (London: Conway Maritime Press, 1981), p. 128.
10. Brian Lavery and Simon Stephens, *Ship Models* (London: Philip Wilson, 1995), p. 58.
11. Samuel Pepys, *The Diary of Samuel Pepys*, Henry Wheatley ed, Vol. 2 (London and New York: George Bell and Sons, 1903), p. 286.
12. Pepys, p. 326.
13. John Campbell, *Lives of the British Admirals*, Vol. 2 (London: Printed for C J Barrington, 1812), p. 514.
14. Frank Fox has recently concluded that the model at Trinity House, London, long considered to represent the *Loyal London* on the basis of the carved arms of Sir Jeremy Smith on the stern and Sir Jeremy's association with the *Loyal London*, is instead a model of the *Royal Charles* of 1668. It is known that Sir Jeremy asked Jonas Shish to make a model for him in 1672, and the *Charles*, which was a Shish product, closely resembles this model.

Chapter 2 A Charles II 5th rate c1680

1. J R Tanner, ed, *Further Correspondence of Samuel Pepys* (London: G Bell and Sons Ltd, 1929), pp. 295–6.
2. Clements R Markham, *Life of Robert Fairfax of Steeton* (London: Macmillan and Co., 1885), p. 23.

Chapter 3 The *Coronation*, 2nd rate of 1685

1. Henry Culver, *Contemporary Scale Models of Vessels of the Seventeenth Century* (New York: Payson & Clarke Ltd, 1926), p. 50.
2. R C Anderson, personal correspondence, 15 July 1975.
3. H Percy Ashley, 'H.M.S. Coronation', *Shipmodeler* (September 1930), p. 162.
4. R C Anderson, 'Books', *The Mariner's Mirror*, 17 (1931), p. 420.
5. Samuel Pepys, *Memoirs Relating to the State of the Royal Navy of England, for Ten Years, Determined December 1688* (London: Printed for Ben Griffin, 1690), p. 62.
6. C Dibdin, *Sea Songs and Ballads* (London: Ticknor & Fields, 1866), pp. 64–6.

7. Peter Watson, 'The Booty off Penlee Point', *The Sunday Times* (20 November 1977), p. 13.

Chapter 5 **A William III 4th rate c1695**
1. John Entick, *Naval History* (London: Printed for R Manby, W Reeve, W Bizet, P Davey, B Law, & J Scott, 1757), pp. 499–500.

Chapter 6 **The *Northumberland*, 3rd rate of 1702**
1. J E Parry, et al, *Export of Works of Art 1983–84* (London: Her Majesty's Stationery Office, 1985), p. 66.
2. Statement of Expert Adviser, Head of Department of Ships, National Maritime Museum, to Reviewing Committee on the Export of Works of Art, August, 1983.
3. Franklin, John, *Navy Board Ship Models 1650–1750* (London: Conway Maritime Press Ltd, 1989), p. 131.

Chapter 7 **The *Marlborough*, 2nd rate of 1706**
1. R C Anderson, *The Rigging of Ships in the Days of the Spritsail Topmast 1600–1720* (Salem: Marine Research Society Salem, 1927).
2. W G Perrin, contrib, 'The Ornamentation of Men of War 1703', *Mariner's Mirror*, 3, 1 (1913): pp. 20–21.

Chapter 8 **The *Diamond*, 4th rate of 1708**
1. R Morton Nance, *Sailing-Ship Models* (London: Halton & Truscott Smith Ltd, 1924), pp. 73–4, plates 64, 65.
2. John Preble, *The Darien Disaster* (New York: Holt, Rinehart and Winston, 1969).
3. Preble, p. 12, 21.
4. Preble, p. 301.

Chapter 9 **A Queen Anne 3rd rate c1710**
1. Philip Macdougall, *Royal Dockyards* (London: David & Charles Limited, 1982), pp. 41–42, 216.

Chapter 10 **The *Royal Oak*, 3rd rate of 1713**
1. T Sutherland, engraver, 'The Queen's Drawing Room, Windsor Castle', W H Pyne, 1 August 1816.
2. Reprinted in Matthews, William, ed, *Charles II's Escape From Worcester* (Berkeley: University of California Press, 1966).

Chapter 11 **The *Diamond*, 5th rate of 1723, and the *Greyhound*, 6th rate of 1720**
1. Frank R Stockton, *Buccaneers and Pirates of Our Coasts* (Macmillan Books, 1897), pp. 275–6.
2. The 'Great' Edward Low, 'The Most Merciless Pirate Known to Modern Times', *The New York Times*, 14 August 1892, p. 12.
3. John F Watson, *Annals of Philadelphia* (Philadelphia: E L Carey & A Hart, 1830), p. 469.
4. Rex Whitworth, *William Augustus Duke of Cumberland: A Life* (London: Leo Cooper, 1992), p. 25.

Chapter 12 **The *Lion*, 4th rate c1738**
1. 'Ships' Models', *Antique Collecting*, 8, 1 (May 1973), p. 24.
2. John Stalker and George Parker, *A Treatise on Japaning and Varnishing* (Oxford: John Stalker and George Parker, 1688), reprinted, H D Molesworth, ed (Chicago: Quadrangle Books, 1960).

Chapter 13 **A George II 4th rate c1745**
1. 'Fine Historical Steam Engine Models and Ship Models', Christie, Manson & Woods, 17 November 1971, lot 155.

Chapter 14 **The *Namur*, 3rd rate of 1746**
1. Wm Laird Clowes, *The Royal Navy a History From the Earliest Times to the Present*, Vol. 3 (London: Sampson, Low, Marston and Company, 1898), p. 51.

Chapter 15 **A French 64-gun ship c1754, built by Augustin Pic**
1. National Archives, Paris, Navy Holdings, file D2 34.
2. National Archives, Paris, Navy file C7246 Personnel Records.

Chapter 16 **The *Généreux*, 3rd rate of 1785**
1. Manly Plaindealer, *The Wooden World Dissected*, 2nd edition (London: H Meere, 1708).

Chapter 17 **The *Franklin*, American 74-gun ship c1800**
1. Papers of the War Department 1784–1800, Roy Rosenzweig Center for History and New Media.
2. Joshua Humphrey's papers, Collection 0306, The Historical Society of Pennsylvania.

Chapter 21 **A Queen Anne royal barge**
1. For accounts of the wreck of the *Association*, see: Larn, Richard, *Shipwrecks of the Isles of Scilly* (Cornwall: Troutbeck Press, 1999).
2. Geoffrey Callender, *Sea Kings of Britain* (London: Longmans, Green and Co., 1939), p. 112.

Chapter 22 **An admiral's barge c1710**
1. Thomas Lediard, *The Naval History of England* (London: John Wilcox and Oliver Payne, 1735), p. 589.
2. Samuel Pepys, *The Diary of Samuel Pepys*, Henry Wheatley, ed, Vol. 2 (London and New York: George Bell and Sons, 1901), p. 345.
3. George Berkley, *The Naval History of Britain* (London: T Osborne and J Shipton, 1756), p. 509.
4. Harm Stevens, *An English Taffrail*, in *The Art of Technology*, Harm Stevens, ed (Wormer: Inmere BV, 1995), pp. 13–15.
5. Samuel Pepys, *The Diary of Samuel Pepys*, Henry Wheatley, ed, Vol. 2 (London and New York: George Bell and Sons, 1901), p. 356.
6. Andrew Marvell, *Last Instructions to a Painter, London 1667*, in *Andrew Marvell: The Complete Poems*, Elizabeth Story Donno, ed (New York: Penguin, 1972).

Chapter 23 **A Georgian admiral's barge**
1. Phineas Pett, *The Autobiography of Phineas Pett*, W G Perrin, ed (London: Navy Records Society, 1918).
2. Peter Munday, *The Travels of Peter Mundy in Europe and Asia, 1608–1667*, Lt Col Sir Richard Carnac Temple, Bart, ed, Second Series, No. 45 (London: Hakluyt Society, 1919).
3. A W J, 'Official Ship Models', *Mariner's Mirror*, 14 (1928), p. 282.
4. L G Carr Laughton, 'Official Ship Models', *Mariner's Mirror*, 13 (1927), p. 175.
5. Jules Sottas, 'Naval Museums III. France', *Mariner's Mirror*, 5 (1915), p. 77.
6. Samuel Pepys, *The Diary of Samuel Pepys*, Henry Wheatley, ed, Vol. 3 (London and New York: George Bell and Sons, 1901), pp. 368–369.
7. David Green, *Queen Anne* (New York: Charles Scribner's Sons, 1970).
8. H Szymanski, 'Naval Museums-VI., Germany', *Mariner's Mirror*, 8 (1922), p. 337.
9. John Franklin, *Navy Board Ship Models 1650–1750* (London: Conway Maritime Press, 1989), p. 90.
10. Donald McNarry, *Ship Models in Miniature* (Newton Abbot: David & Charles, 1975), p. 68.
11. Angus White, personal communication.
12. Angus White and Laurie Langford.
13. Angus White, personal communication.
14. Brian Lavery and Simon Stephens, *Ship Models: Their Purpose and Development from 1650 to the Present* (London: Philip Wilson, 1995), 256, see p. 39.
15. NMM, ref. no. SLR0408, 1719-2, formerly in the Royal Naval Museum.

Chapter 24 **An admiral's barge c1775**
1. We are grateful to Amy Miller of the NMM, who provided us with information concerning naval dress in the eighteenth century.

Chapter 25 **A troop transport c1810**
1. We are indebted to Major Grant Walker, who kindly provided information concerning the uniforms of the sailors and soldiers and identified them.

Chapter 26 **A ship's boat c1750**
1. William Bligh, *A Narrative of the Mutiny on Board His Majesty's Ship Bounty* (London: G Nicol, 1790).

Chapter 27 **A Dutch state yacht c1690**
1. We are grateful to Hendrik Busmann for his scholarly suggestions concerning the possible meaning behind these ship carvings.
2. Julius Wilhelm Zincgreff, *Emblematum ethico-politicorum centuria Julii Guilielmi Zincgrefii*, Coelo Matth, Meriani, 1619, Prostat apud Johann Theodor de Bry [Heidelberg].

Chapter 29 **Model of the *Queen Charlotte* figurehead c1784**
1. Carr Laughton, L G, *Old Ship Figureheads and Sterns* (Halton & Truscott Smith Ltd, New York, 1925), pp. 81–83.
2. McCarthy, Erica, 'Ship Carvers in Eighteenth and Nineteenth Century Britain', *Sculpture Journal*, 24.2 (Liverpool University Press, 2015), p. 188.

Chapter 30 **Model of the foremast of the *Victory* of 1765 with damage sustained at Trafalgar**

1. Leetham, Arthur, *Official Catalogue of the United Service Museum* (Southwark: J J Keliher & Co, 4th edition, 1914), p. 88d.
2. Alexander M Bradley and Richard G Bartelot, *The Three Dorset Captains at Trafalgar* (J Murray, London, 1906), Appendix B.

Chapter 31 **Ship models in perspective painted on panels**

1. John Entick, *A New Naval History* (London, printed for R Manby, 1757), p. 799.
2. M Witt, 'A Picture of the *Royal George* of 1756', *The Mariner's Mirror*, 39, 1 (February 1953), pp. 59–60

Chapter 32 **Photojournalism in the seventeenth century**

1. *Oud Holland*, 18 (1900), p. 38.
2. ARA, *Archieven der Admiraliteitscolleges*, Vol. XLVII No. 4, Den Haag.
3. Van de Velde, *Van de Velde Drawings in the National Maritime Museum*, Vol. 1 (Cambridge: Cambridge University Press, 1973), p. 342.
4. Public Records Office, London, February 1673/4, King's Bills, S07/40.
5. Public Records Office, London, Minutes of Board, Admiralty, 3/10.

Chapter 33 **Napoleonic-era ship models**

1. Ewart Freeston, *Prisoner-of-War Ship Models* (Annapolis, Maryland: Naval Institute Press, 1973).
2. Clive Lloyd, *A History of Napoleonic and American Prisoners of War 1756–1816: Hulk, Depot and Parole* (Woodbridge, Suffolk, UK: ACC Art Books).
3. Jane Toller, *Prisoners-of-War Work 1756–1815* (Cambridge: The Golden Head Press, 1965), p. 2.
4. Olive Milne Rae, 'Old Dieppe Ivories', *The Connoisseur*, Vol. 21 No. 83, July 1908, pp. 160–166.
5. Culver, Henry, 'Old Prison-Made Bone Ships', *Antiques*, Vol. 8 No. 3, Sept 1925, p. 146.

Appendix **A William III 4th rate c1695**

1. Culver, Henry, *Contemporary Scale Models of Vessels of the Seventeenth Century* (New York: Payson & Clarke Ltd, 1926), p. 25.

Index

Aardenburg 221–22
  admiral's barge (c1710) 194–99
    acquisition 195
    condition 196
    construction 196–7
    exhibitions 198
    historical perspective 198–9
    literature 197
    photographs/illustrations *194, 195, 196, 197, 198, 199*
    provenance 195
  admiral's barge (c1775) 206–9
    condition 207
    exhibitions 208
    historical perspective 208
    photographs *206, 207, 208, 209*
    provenance 207
Adventure (1646) 202
Adventure, 5th rate (1691) 60–65
  acquisition 61
  condition 61
  construction 62
  historical perspective 64–5
  literature 63
  photographs *60, 61, 62, 63, 64, 65*
  provenance 61
Aetna bomb (1776) 180–83, 187
  acquisition 181
  condition 181
  construction 181–2
  historical perspective 183
  literature 182
  photographs *180, 182, 183*
  provenance 181
Alexander 167
Anderson, R C 13, 15, 55, 87, 94
  The Rigging of Ships in the Days of the Spritsail Topmast 13, 55, 87
Anglo-Dutch Wars 35, 98, 258
  first 253
  second 32, 35, 198, 199, 253, 254
  third 33, 34, 36, 58, 64, 72, 256
Antique Collecting 131
Ashley, Captain H Percy: *A Noteworthy Shipmodel* 55
Association 191–3
Assurance 202

Bien-Aimé (French) 160
Blackwall shipyard 6, 90, 96
Blenheim 88, *89*
Bligh, William 216–17
Bonaventure 46
Bounty 216–17
Boyne 128, 203, 204
Britannia (1719) 112, 225, *263*

Caird, Sir James 14, 16
Cape Finisterre, Battle of (1747) 138, 145
Capricieux (France) 160
Carcass bomb (1758) 176–9
  acquisition 177
  condition 177
  construction 177–8
  historical perspective 179
  photographs/illustrations *176, 177, 178, 179*
  provenance 177
care and conservation 275
Carr Laughton, L G: *Old Ship Figureheads and Sterns* 232
Charles II 5th rate (c1680) 14, 38–47
  acquisition 39–41
  condition 41
  construction 41–2
  historical perspective 43–6
  literature 42
  photographs/illustrations *17, 38, 39, 40, 41, 42, 43, 44, 45, 46, 47*
  provenance 41
Christ's Hospital Mathematical School 23, 203
Churchill, John (Duke of Marlborough) 87, 88, 89
Cleveley, John the Elder 108, 227
Collins Jr, Clarkson A 14, 67–8, 211
Columbus (USA) 175
Content (French) 65
Copenhagen, Battle of (1801) 186–7, 213
Coronation, 2nd rate (1685) 48–59, 85, 87
  acquisition 49–52
  condition 55
  construction 55–7
  exhibitions 57
  historical perspective 58–9
  literature 57
  photographs *48, 49, 50, 51, 52, 53, 54, 56, 59*
  provenance 52–4
Culver, Henry 273, 279,
  Contemporary Scale Models of Vessels of the 17th Century 24, 52, 55, 61, 67, 68

Darien scheme 97, 98–9
Deane, Sir Anthony 13, 23, 31, 32–3, 203, 262
Doctrine of Naval Architecture 28, 32
Dentelle (French) 136

Deptford dockyard 46, 64, 108, 116, 126, 136, 202, 203, 227, 257
Devonshire 145, 146
Diamond, 4th rate of 1708 *16*, 90, 91, 92–101
  acquisition 93–4
  condition 94
  construction 94–6
  exhibitions 96
  historical perspective 96–100
  literature 96
  photographs *92, 93, 94, 95, 97, 99, 100, 101, 274*
  provenance 94
Diamond, 5th rate (1723), and *Greyhound*, 6th rate (1720) 122–9
  acquisition 123
  provenance 123
  description 123
  condition 124
  construction 124–5
  exhibitions 125
  literature 125
  historical perspective 126–9
  photographs/illustrations *122, 123, 124, 125, 126, 127, 129*
Diligente (French) 65
Dilkes, Captain Thomas 64–5
Dolfijn (Dutch) 35
Dragon 64
drawings of ship models from the 18th century 250–52
  Boreas (Dutch) *250*
  De Liefde (Dutch) *252*
  Royal William 251
  unidentified eighteenth-century three-decker 251
  Victory (1737) *250*
Dutch state yacht (c1690) 218–23
  acquisition 219
  condition 219
  construction 219–20
  exhibitions 221
  historical perspective 221–22
  photographs *218, 219, 220, 221, 222, 223*
  provenance 219

Eagle 191
Ecureuil (French) 138
Edgar 82, *108*
Edward Battine: *The Method of Building, Rigging, Apparelling and furnishing his Majesties Shipps of Warr according to their Rates* 24, 25, 27, 28
Elisabeth (French) 136

England's Great Loss by a Storm of Wind 58
Éveillé (French) 160
Explosion 186
export licences 21, 23, 76–79, 279

Fairfax, Admiral Robert 41, 43, 44, 46
fakes and forgeries 276–8
figureheads 42, 169, 175, 220, 221, 276
  Adventure 61, *61*
  Charles II 5th rate *17*, 42, *42*, 46
  Coronation 52, 56, 57
  Diamond 94, *94*
  Dutch state yacht *220*, 221
  French gunship *152*, 153, 154
  Lion 133
  Marlborough 86, 87, 88, *89*, 90
  Namur 17, 148
  Northumberland 80, *82*
  photographs/illustrations *17, 42, 220, 251, 266, 270, 271*
  Queen Anne 3rd rate *104*
  Queen Charlotte 228–37
  Royal James 19, 24, 26, 28
  Royal Caroline 224–27
  William III 4th rate 69, 72, 280, *282*
Firebrand 191
foremast of the *Victory* (1765) with damage sustained at Trafalgar 238–43
  acquisition 239–40
  condition 241
  construction 241
  exhibitions 240
  historical perspective 241–3
  literature 240
  photographs/illustrations *238, 239, 240, 241, 242, 243*
  provenance 240
Foudroyant 167
Fox, Frank 31
  Great Ships: the Battlefleet of Charles II 24
Franklin, American 74-gun ship (c1800) 168–76
  acquisition 169
  condition 169–70
  construction 170–71
  historical perspective 172–5
  photographs/illustrations *168, 169, 170, 171, 172, 173, 174, 175, 176*
  provenance 169
French 64-gun ship (c1754), built by Augustin Pic 150–61
  acquisition 151

  condition 152–3
  construction 154–5
  exhibitions 156
  historical perspective 157–60
  literature 155
  photographs/illustrations *150, 151, 152, 153, 154, 155, 156, 157, 158, 159, 161*
  provenance 151–2

Généreux, 3rd rate (1785) 23, 162–7
  acquisition 163
  condition 163
  construction 163–5
  exhibitions 165
  historical perspective 166–7
  literature 165
  photographs/illustrations *162, 163, 164, 165, 166, 167*
  provenance 163
George II 4th rate (c1745) 140–43
  acquisition 141
  condition 141
  construction 141
  historical perspective 141–2
  photographs *140, 141, 142, 143*
  provenance 141
Georgian admiral's barge (c1720) 200–205
  acquisition 201
  condition 201
  construction 201–2
  historical perspective 202–5
  photographs *200, 201, 203, 205*
  provenance 201
Glorieux (French) 160
Glorious First of June, Battle of the (1794) 229, 235
Golden Lion (Dutch) 72
Groot Hollandia (Dutch) 35

Hampton Court 203
Harrison, John 193
Harwich 58
Hecla 186
Herbert, Thomas (Earl of Pembroke) 75, 80, 82
History of the British Navy 15
Hoare, Captain Charles 14, 15
Humphreys, Joshua 168–75

identifying ships at sea 72
Illustrated London Times 15
Independence (USA) 175
Inflexible (French) 160

Jacobin (French) 235
japanning 112, 138

287

Jennings, Admiral Sir John 195, *195*
John Campbell: *Lives of the British Admirals* 36
Johnson, William 6, 90, 96

Kent, Sir Richard Gough 52, 87

L'Artésien (French) 160
L'Assuré (French) 160
Leander 166–7
Lediard, Thomas: *The Naval History of England* 197
Lenox 193
Lion (1710) 106
Lion, 4th rate (c1738) 128, 130–39
   acquisition 131–2
   condition 133
   construction 133–4
   exhibitions 135
   historical perspective 136–8
   literature 134–5
   photographs/illustrations *130, 131, 132, 134, 135, 136, 137, 139*
   provenance 132–3
Lizard 82, 204
London Times, The 15
Loyal London 198

Mariner's Mirror, The 15, 19, 24, 55
Marlborough, 2nd rate (1706) 6, *16*, 54, 84–91
   acquisition 85–7
   condition 87
   construction 87–9
   exhibitions 89
   historical perspective 90–91
   literature 89
   photographs *84, 85, 86, 87, 88, 89, 91*
   provenance 87
models in perspective painted on panels 245–50
   frigate *Enterprise* (1776) 248, 249
   Hector, 3rd rate (1774) *247*, 249
   literature 250
   Royal George, 1st rate (1756) *247*, 249
   Royal William, 1st rate (1719) *244*, 248
   unidentified frigate (c1770) *248*, 249
   unidentified frigate (c1770) 249, *249*
   Valiant 74 guns (1759) 249, *249*
   Victory, 1st rate (1738) *246*, 249
Montagne (French) 235
Montagu, Edward (Earl of Sandwich) 34, 35, 36
Morduant 46
Morgan, Junius Jr 15, 51, 54, 61, 85, 87, 145
Mystic Seaport Museum 16, 40, 67, 68, 71, 211, 212

Namur, 3rd rate (1746) *17*, 144–9
   acquisition 145
   condition 145
   construction 145–7
   historical perspective 147–8
   literature 147
   photographs *144, 145, 146, 147, 148, 149*
   provenance 145
   War of Jenkins's Ear 147–8
Nance, R Morton 14
   Sailing-Ship Models 93
Napoleonic Wars 183, 212–13, 265–73
Napoleonic-era ship models 264–74
   characteristics 270–71
   construction 270
   historical perspective 272–3
   legacy 273
   origins 265–9
   photographs/illustrations *264, 265, 266, 267, 268, 269, 270, 271, 272, 273, 274*
   tools 272
Nautical Research Journal 13
Nelson, Horatio 36, 166, 177, 179, 187, 239, 240, 241–2
NMM (British National Maritime Museum) 16, 31, 77–80, 85, 106, 181, 203, 204, 215–16
   Britannia 112
   Caird, Sir James 14
   Morduant 46
   R C Anderson 13, 15
   Royal George 249
   Royal James 23
   Royal William 133
   Saint Michael 36
   Stephens, Simon 21, 23, 81
   Victory 246
Nonsuch 202
Northumberland, 3rd rate (1702) 58, 74–83, 167
   acquisition 75–80
   condition 80
   construction 80–81
   exhibitions 81
   historical perspective 82–3
   literature 81
   photographs *74, 75, 76, 77, 79, 80, 82, 83*
   provenance 80
Nuestra Senora del Rosario (Spanish) 116

origins of yachting 222

Pallas 49
Pepys, Samuel 15, 24, 58, 61, 78, 117, 118
   diary 198, 199, 203
   Edward Battine 28
   letters 43
   Naval Minutes 14, 23
   Secretary of the Navy Board 13, 28
   Sir Anthony Deane 23, 32, 33, 203
   William Coventry 13, 203

Pett, Christopher 13, 203, 256
Pett, Peter 199, 202, 203, 261
Pett, Phineas 43, 202, 263
Phoenix 32
photojournalism in the 17th century 253–63
   Assistance 257
   Britannia 263
   Caleb 255
   Gouda (Dutch) 259
   Leeuwarden (Dutch) 258
   Richard 256
   Royal Charles 262
   Royal James 262
   Sovereign of the Seas 261
   St Andrew 260
   Triumph 253
   Vereenigde Provincien (Dutch) 257
Pic, Augustin 150–61
pissdales *16*, 96, 105, 114, 115
Plymouth 65
pop-up models 123, 124, 125, 127–8
Prince 28, 32, 36, 57, 72
Prince of Parma (French) 160
Prince Royal 108, 202
Pro Patria (Dutch) 198

Queen Anne 3rd rate (c1710) *16*, 102–9
   acquisition 103
   condition 104
   construction 104–5
   historical perspective 106–8
   literature 105
   photographs/illustrations *102, 103, 104, 105, 106, 107, 108, 109*
   provenance 103–4
Queen Anne royal barge 188–93
   acquisition 189, 211
   condition 189
   construction 189–91
   historical perspective 191–3
   literature 191
   photographs/illustrations *188, 189, 190, 191, 192, 193*
   provenance 189
Queen Charlotte 234–6
Queen Charlotte figurehead (c1784) 228–37
   acquisition 229
   condition 230–31
   construction 231
   exhibitions 232
   historical perspective 232–6
   literature 232
   photographs/illustrations *228, 229, 230, 231, 232, 233, 234, 235, 236, 237*
   provenance 229–30

Racehorse 179
Restoration 82
restrictions on carving 74, 82, 90–91, 96, 105
Revenge 16, 112
Rippon 106
Rogers, Colonel Henry Huddleston 15, 51, 67, 68, 93–4, 248

Romney 191
Royal Caroline 227, *227*
Royal Caroline figurehead 1750 224–7
   acquisition 225
   description 226–7
   photographs/illustrations *224, 225, 226, 227*
Royal Charles 31, 32, 36, 72, 117, 198, 199, *262*,
Royal George 23, 36, 112, 236, 239, *242*, *243*, 247, 249
Royal James, 1st rate (1671) *18–37*, 72, 203, 256, 262
   acquisition 19–23
   condition 24–28
   construction 28–31
   exhibitions 31
   historical perspective 32–6
   literature 31
   photographs *18, 20, 21, 22, 26, 27, 29, 30, 33, 34, 37*
   provenance 23–4
Royal Oak, 3rd rate (1713) 16, 23, 58, 110–121, 198
   acquisition 111
   condition 112
   construction 112–115
   exhibitions 116
   historical perspective 116–120
   literature 115
   photographs/illustrations *16, 110, 111, 112, 113, 114, 115, 116, 117, 119, 121*
   provenance 111
Royal Sovereign 16, 72, 90, 204, 242
Royal William 133
Rupert 32, 65
RUSI (Royal United Service Institute) Museum 19, 20, 24, 123, 131, 132

Science Museum, London 16, 28, 36, 57, 75, 78, 131, 204, 246, 247
Scots Guards 213
Seahorse 179
Sergison, Charles 15, 94
   collection 15, 62, 93
ship's boat (c1750) 214–17
   acquisition 215
   condition 215
   construction 215–16
   historical perspective 216–17
   photographs *214, 216, 217*
   provenance 215
Shipmodeler, The 55
Shortiss, Robert 82, 83
Shovell, Sir Cloudesley 191–3
Society for Nautical Research 13, 15, 16
Solebay, Battle of (1672) 32, 33, 34¬6, 254, 255, 256
Spence, Robert 15, 20, 21, 23, 24, 61
St Albans 21, 61, 203, 204
St George 72, 191
St Michael 36, 90, 276
Strombolo 186
Success 167
Sulphur bomb (1797) 184–7

   acquisition 185
   condition 185
   construction 185
   historical perspective 185–7
   literature 185
   photographs *184, 185, 186, 187*
   provenance 185
Surprise 120

Tartaru 186
Terror 183
The Wooden World Dissected 166
Thunder 183
Trafalgar, Battle of (1805) 36, 238–43
trenails 88, 169, 190
Trident 65
troop transport (c1810) 210–13
   acquisition 211
   condition 211
   construction 211
   historical perspective 212–13
   photographs *210, 211, 212, 213*
   provenance 211

uniforms *198*, 208–9, 211, 212–13
Union Jack 72, 95, 99, 196

van de Velde, Willem (elder and younger) 31, 32, 253–62
Vaughn, John (Earl of Carbery) 52, 87
Vesuvius 183
Victory (1691, 1715) 36
Victory (1737) 36, 128, 136, 245, 246, *246*, 249, *250*
Victory (1765) 36, 233, 238–43
Volcano 186

War of Jenkins's Ear 147–8
Warwick 160
Washington (USA) 175
Waverley criteria 77–8
William III 4th rate c1695 (appendix) 278–82
   acquisition 279
   condition 280
   construction 281
   photographs *278, 279, 280, 281, 282*
William III 4th rate (c1695) 66–73
   acquisition 67–8
   condition 68
   construction 68–71
   exhibitions 71
   historical perspective 72
   literature 71
   photographs *66, 67, 68, 69, 70, 71, 73*
   provenance 68
Williams, Guy R: *The World of Model Ships and Boats* 103
Woolwich dockyard 13, 23, 32, 108, 112, 116, 202, 204

Zephyr (French) 138